Science of the Seance

Transnational Networks and Gendered Bodies in the Study of Psychic Phenomena, 1918–40

BETH A. ROBERTSON

UBC Press • Vancouver • Toronto

25 24 23 22 21 20 19 18 17 16   5 4 3 2 1

Printed in Canada on FSC-certified ancient-forest-free paper (100% post-consumer recycled) that is processed chlorine- and acid-free.

Library and Archives Canada Cataloguing in Publication

Robertson, Beth A. (Beth Anne), author
Science of the seance : transnational networks and gendered bodies in the study of psychic phenomena, 1918–40 / Beth A. Robertson.

Includes bibliographical references and index.
Issued in print and electronic formats.
ISBN 978-0-7748-3349-3 (hardback).–ISBN 978-0-7748-3350-9 (pbk.).–
ISBN 978-0-7748-3351-6 (pdf).–ISBN 978-0-7748-3352-3 (epub).–
ISBN 978-0-7748-3353-0 (mobi)

1. Parapsychology–Research–History–20th century. 2. Spiritualism–Research–History–20th century. 3. Parapsychology and science–History–20th century. 4. Science and spiritualism–History–20th century. 5. Parapsychologists–History–20th century. 6. Sex role. I. Title.

BF1028.R63 2016   130.72′2   C2016-905163-3
C2016-905164-1

Canada

UBC Press gratefully acknowledges the financial support for our publishing program of the Government of Canada (through the Canada Book Fund), the Canada Council for the Arts, and the British Columbia Arts Council.

This book has been published with the help of a grant from the Canadian Federation for the Humanities and Social Sciences, through the Awards to Scholarly Publications Program, using funds provided by the Social Sciences and Humanities Research Council of Canada.

Printed and bound in Canada by Friesens
Set in Garamond by Apex CoVantage, LLC
Copy editor: Robert Lewis
Cover designer: TG Design
Proofreader: Alison Strobel

UBC Press
The University of British Columbia
2029 West Mall
Vancouver, BC V6T 1Z2
www.ubcpress.ca

Contents

Illustrations

Acknowledgments

The trajectory that brought me to study interwar psychical research was long and varied – a history that now haunts this work. I was first introduced to some of the amazing images that this book features in 2004 by James Opp at Carleton University. He perhaps deserves the most thanks, as it was his unfailing support and attentive critique that made this work what it is today. Carleton University was in general an extraordinary place to undertake this research, and many provided guidance, encouragement, and advice along the way. Thank you especially to Joanna Dean, Jennifer Evans, Patrizia Gentile, Michel Hogue, Andrew Johnston, Dominique Marshall, James Miller, Alexis Shotwell, and John Walsh. Competent administrative staff, including Regina Aulinskas, Irene Sanna, and Joan White played a key role. Others outside Carleton deserve thanks as well, including Michel Ducharme, Pamela Klassen, Bertrand and Jack Dufresne, Edward Jones-Imhotep, Robert MacDougall, Gillian McCann, Molly McGarry, and James Moran.

I should also thank the diligent staff of the University of Manitoba Archives and Special Collections, especially Brian Hubner, Andrea Martin, and Shelley Sweeney. Also deserving much gratitude are Jane Britton, Jessica Blackwell, and Nick Richbell at the University of Waterloo Library, Claire Hatty and Grant Young at Cambridge University Library, and Jeff Twine at the American Society for Psychical Research.

My editor at UBC Press, Darcy Cullen, has been truly amazing and went far beyond the call of duty on several occasions. To her and the rest of the skilled staff of UBC Press, thanks so much. Thanks also to two anonymous external reviewers, who ensured the scholarship of this study was sound and provided clear and refreshing insight.

In addition, I would like to thank some dear friends – vastly intelligent and gifted individuals who have been there to listen, make me laugh, and offer fresh perspectives. These wonderful people include Susan Joudrey, Karen and Brian Foster, Tamara Krawchenko, Tyler Knowlton, Maureen Mahoney, Madelaine Morrison, Justin Rivest, Kathryn Desplanque, Amanda Sauermann, Lloyd Solomon, and Jonathan Collicott. A special thank you must also go to Laura Madokoro, who, in addition to being a wonderful friend and inspiration, offered invaluable advice. Yet one more person who has been a most ardent supporter is Glendon McKinney. Thank you so much for your book recommendations, inspiring poetry, and wonderful distractions along the way.

Last, I must thank two very cherished individuals. The first is William Knight. He has consistently tolerated the long hours, late nights, and odd habits to which my writing has led to. The second is Tamar Joan Priscilla. She tumbled into the world while I wrote and refined this book. Since then, she has continued to be a source of hope and love while inspiring me in ways I cannot fully express.

Science of the Seance

Groping in the Dark: An Introduction

> This world is a laboratory of souls, a forcing ground where the material refines out the spiritual.
> – Arthur Conan Doyle, *The History of Spiritualism*[1]

The subject of this study remains wedged between discourses of science and religion, matter and mind, materialism and metaphysics. Examining a twentieth-century network of psychical researchers, mediums, and spirits, I argue that the study of paranormal phenomena was not some bizarre digression in a progressively secular world. Rather, experimentation with the limits of death was intimately connected to the culture of its era. Interest in the paranormal has often been relegated to the nineteenth century, a supposition that is buttressed by a wealth of literature dedicated to Victorian spiritualism. From Marlene Tromp and Molly McGarry to Pamela Thurschwell and many more, these authors have demonstrated that spirit seeking was formative of nineteenth-century society.[2] Yet the practice of investigating what lies beyond the grave did not diminish but gained momentum following the outbreak of the First World War. Situated along the fault lines of science and spirituality, matter and the psyche, the twentieth-century seance was much more than a continuation of nineteenth-century preoccupations. Just as empiricism, technology, perception, gender, class, race, and the body were being reconstructed in this period, so too did the seance experience a radical shift.

This book contends that interwar psychical researchers, mediums, and spirits did not simply mimic but engaged with and ultimately transformed broader cultural conceptions of empiricism and technological praxis, as well as gendered, class, and racial power within the confines of the seance. Investments in particular kinds of embodiment remained key to such reconfigurations, providing the means by which interwar ideals of scientific authority could be affirmed, as well as displaced and disassembled. Investigators laboured to construct a science of the paranormal by reformulating spiritual encounters into measurable and discernible traces. Representing themselves as scientific and rational, psychical researchers shaped the practice of communing with the dead into an empirical exercise. The dark seance room had been a space in which the earthly encountered the supernatural to reveal another plane of existence. In the hands of twentieth-century psychical researchers, however, the seance became a laboratory of rigorous experimentation. Investigators honed their own senses, used technology, and employed psychological and medical methods. They positioned mediums and ghosts as the embodied evidence of preternatural realities. In the process, investigators, psychics, and spirits became entangled in a gendered and sexual politics that dictated who was to have power over whom. Although researchers endeavoured to maintain a sense of masculine authority, mediums and spirits did not always comply with their experimenters, blurring the lines between scientist and subject, male and female, body and spirit. Refusing to be passive bystanders of the scientific project as it unfolded in the interwar seance room, mediums and spirits wrestled with their investigators to reshape the contours of empiricism. Their participation marked a fundamental breakdown of cultural authority as it was defined within interwar society, their embodied performances reflecting back an image of phallic prowess that undermined its ability to wield power in the first place.

The Unravelled Knot: Transnational Networks, Scientific Endeavours, and Psychical Research

This study focuses on a network of psychical investigators that extended across Canada, the United States, and Britain from 1918 to 1940. Like other transnational histories, it aims "to move beyond a national framework of analysis" in order "to explore connections between peoples, societies and events usually thought of as distinct and separate."[3] Although rooted in distinct national and geographical contexts, these individuals maintained their associations across substantial distances, sharing methods, technologies, mediums, and even spirits. Whether in New York or Boston, in

St. Catharines, Ontario, or Winnipeg, Manitoba, or in London, England, these investigators and the mediums they studied significantly contributed to the project of psychical research in the interwar era.

Some of the researchers and mediums of this network have been studied individually by scholars. What have been overlooked are the links that existed between them. Rather than viewing these investigations through a strictly nationalist lens, and thus regarding them as isolated and disconnected, I draw these sources together. Although they were not the only ones probing the spirits at this time, I have focused on these particular researchers and mediums because they demonstrated the intricate, transnational patterns of communication and collaboration that developed around the pursuit of the paranormal. Through such a network, psychical researchers and mediums legitimized their pursuits as they established alliances across Canada, the United States, and Britain.

This network included an extraordinary medium, Mina Stinson Crandon, or "Margery," as she became known. She originally hailed from Canada but was living in Boston when she first discovered her gifts. Le Roi Crandon, her second husband and a respected surgeon of the city, began examining his wife's uncanny powers in the early 1920s, along with a tight group of individuals. As Margery became internationally recognized, this group expanded to include many more. Investigated by numerous professionals from across the globe throughout the 1920s and 1930s, Margery fostered as much wonder as she did controversy, drawing together psychical researchers intent on uncovering the mysteries of the otherworld. Members from the Society of Psychical Research (SPR) and other institutions in Britain were some of those who sought to examine her gifts. Eric J. Dingwall, the research officer of the SPR, went so far as to pursue Margery across the Atlantic Ocean to prompt a series of strange materializations. The American Society for Psychical Research (ASPR) similarly showed significant interest in Margery's psychic capacities, and several of the society's members across North America requested a sitting with her. They included physician T. Glen Hamilton and his wife, Lillian, of Winnipeg.

The Hamiltons conducted some of the most elaborate photographic experiments of paranormal phenomena in the world in the 1920s and 1930s, leading T. Glen Hamilton to make significant contributions to the *Proceedings* and the *Journal of the American Society for Psychical Research* and to the *Quarterly Transactions of the British College of Psychic Science,* among other publications.[4] Joining the fray of internationally renowned psychical researchers, Hamilton allied himself with numerous investigators in the United States and Britain. One of his most significant collaborations was with the Crandons of Boston. From the mid-1920s until the 1930s, the experiments of the Hamiltons and the Crandons intimately intertwined. In addition to correspondence, Hamilton and Crandon joined forces on several occasions and even shared the ghost of Margery's dead brother, Walter.

Walter revealed himself in surprising places, including at the New York spiritualist retreat of Lily Dale through a medium known as William Cartheuser. This unexpected connection drew the attention of ASPR psychical researchers, who attempted to test and experiment with Cartheuser. None of their investigations, however, were to be as detailed or longstanding as those done by author and musician Jenny O'Hara Pincock – a woman who regularly frequented Lily Dale and eventually convinced Cartheuser to return home with her to St. Catharines, Ontario. Although O'Hara Pincock remained very much on the edge of the scientific pursuit of the paranormal, unabashedly founding the Church of Divine Revelation and the Radiant Healing Centre in the 1930s, her testimony of Cartheuser's mediumship was internationally recognized as "authoritative" and even scientific in its approach.[5]

Drawing together sources from Canada, the United States, and Britain, this book brings to the fore significant qualities of paranormal research in the interwar era. In addition to a broad range of published materials from the period, resources gathered for this study include those from the archives of the British and American Societies for Psychical Research, as well as the Hamilton Family Fonds at the University of Manitoba and the Maines Pincock Family Fonds at the University of Waterloo. Notably, a substantial amount of the archival material addressed in this book emerges from these two Canadian collections. There are compelling reasons for this emphasis. Together, they form a coherent body of materials that has allowed me to identify a host of transnational exchanges between certain centres of paranormal investigation that may have been overlooked otherwise. All the while, these extraordinarily detailed records have provided a unique opportunity to tease out the complexities of the cultural and political work performed by individual psychical researchers and mediums. A deep analysis of these Canadian collections, situated in dialogue with American and British sources, uncovers important yet largely unexamined links between and across national boundaries.

Such a transnational scope is essential to fully understanding how psychical research took shape between the wars. Spiritualists and psychical researchers had communicated across national lines since the nineteenth century.[6] By the 1920s and 1930s, however, the cross-border exchange of mediums, sitters, and researchers had intensified.[7] And they were by no means alone. With radically changing economies, politics, and demographics from 1900 to the 1930s, transnationalism characterized the period. Historian Stephen Kern has argued that the late nineteenth and twentieth centuries witnessed "a new sense of distance" whereby "lines of communication and transportation were extended over unprecedented distances, spreading out and at the same time bringing people into closer proximity than ever before."[8] The First World War only encouraged these connections. From the 1920s onward, transnational organization and collaboration flourished between diverse groups of individuals who were intent on building a "global community," to use the words of historian Akira Iriye.[9] Whether facilitated by religious or humanitarian groups, athletes or intelligentsia, an unprecedented cross-border cooperation marked "the emergence of international society."[10]

In such a milieu, the continuous movement of mediums, sitters, and researchers across borders was far from unusual. Yet this trait has not always been closely analyzed in other studies of interwar spirit seeking. Jenny Hazelgrove, author of *Spiritualism and British Society between the Wars,* for instance, does in fact cite experiments that could be classified as transnational, such as those performed by American researcher Joseph B. Rhine with Irish medium Eileen Garret. Hazelgrove, however, does not highlight this feature of the experiments, maintaining her focus on British culture.[11] My study takes a different approach by probing how such links contributed to wider alliances as psychical researchers across the Atlantic Ocean attempted to construct a science around the paranormal.

These transnational ties were instrumental to how interwar psychical researchers represented themselves as credible scientific professionals, making it possible for them to argue for the empirical legitimacy of their findings across time and space. Arguably, such connections could also have certain disadvantages since reputations of collaborators could fall, as well as rise, in sync. This is important considering that the work of many psychical researchers was questioned and discredited before, during, and well after the interwar period.[12] Yet the network of investigators addressed in this study believed that such risks were necessary to further the science of the paranormal. Their conviction reflects how prominent transnational cooperation became in this era, shaping not only psychical research but also technological innovation, social movements, international diplomacy, economies, and more.[13]

Paranormal investigators strove to obtain the scientific label in a number of other ways as well. Undoubtedly, psychical researchers advocated some unconventional conceptions of scientific, psychological, and philosophical theories, as they seemed to take very literally William James's proposed "radical empiricism."[14] Yet rather than some peripheral pursuit, psychical research demonstrated a great deal about the relationship between the production of empirical knowledge and interwar culture. Historian Steven Shapin has suggested that as opposed to being formed in a vacuum, knowledge is produced by people "against the background of their culture's inherited knowledge, their collectively situated purposes."[15] The significant influence of science on modern society made the label of empiricism "an especially powerful incantation," as several historians have recognized.[16] As a result, a variety of seemingly disparate pursuits struggled to gain such status. Eventually, the label was attained by some, such as evolution and relativity, whereas others were not so successful, including phrenology, psychical research, and the investigation of unidentified flying objects. Rather than demarcating a clear distinction between science and nonscience, a comparison of the genesis of these endeavours reveals that the line between science and its more marginal incarnations is not fixed but is fluid and historically predicated.[17]

The pursuit of the paranormal in the interwar period illustrated this dynamic and often messy interplay between science and culture, empiricism and the esoteric. Spiritualists since the nineteenth century had been convinced that the living could communicate with the dead. Conceptions of life beyond the grave varied widely, but it was uniformly held that the otherworld was another plane of existence in which individual personalities continued to exist. People of this world could re-establish a connection with those who had died by joining hands in dark rooms and calling on the spirits to speak – a practice that came to be called a "seance."[18] Twentieth-century psychical researchers were equally enamoured with the intelligences of the seance room, but they insisted upon scrutinizing discarnate personalities by means of strict tests. Investigators argued that such examinations were necessary to determine whether these entities were the souls of those who had died. Describing themselves as agnostic on the question of the spirits' existence, psychical researchers claimed they were more rational and empirically driven than their spiritualist counterparts. In 1918 the president of the ASPR, James H. Hyslop, explained that spiritualists chose "to remain in the limbo of dark seances and undiscriminating performances, which carry no weight with any intelligent man."[19] He defined psychical researchers, in contrast, as those who employed "the scientific method," remaining skeptical of the spirits' presence until proven, and who pursued the paranormal in the interests of "dignity, real science and intelligent treatment of facts."[20]

Psychical researchers' efforts to distinguish themselves from spiritualists did not convince everyone. *Toronto Star* correspondent Mark H. Halton, for instance, heralded the scientific study of the unseen as refreshingly empirical in comparison to the "many neurotics, ignoramuses and quacks calling themselves spiritualists." Yet, even after questioning British psychical researcher Nandor Fodor, Halton insisted that the investigator "believe[s] you survive after death, and he does believe the dead can communicate with the living. So we'll call him a spiritualist for short."[21] Psychical researchers may have tried to define themselves as objective scientists, yet this did not stop others from interpreting their practices differently. The meaning of psychical research was not transparent and was construed in ways that investigators could not fully control. In this regard, psychical research resembled the natural sciences.

Scholar Mark Erikson describes science as "a complex, contested and contestable family-resemblance concept that holds a range of different meanings according to where it is being deployed and by whom."[22] Unlike psychical research, the multiple understandings and discrepancies of science were carefully smoothed over not only by scientists but also by a society that invested deeply in the scientific project.[23] The characterization of psychical research as a failed science or pseudoscience often resulted in the showcasing of its weaknesses, biases, and inconsistencies.[24] Building off of philosopher and anthropoligist Bruno Latour's metaphor of science as a "very tight knot," I suggest that paranormal investigation was a knot slightly unravelled. Its seams and fractures were therefore easier to discern than were those of the more established sciences.[25] Examining what was eventually perceived to be a failed science affords the opportunity to glimpse what empiricist endeavours look like when they are in the making and thus to discern the tenuous boundary between the scientific and the esoteric.

The ability of psychical research to showcase the inconsistencies of science became troublesome in the early twentieth century, which saw significant challenges to Newtonian causality, resulting in the "blurring of the distinctions between space, time and matter."[26] Such contestations put the institution of science into an uncertain epistemological position, as the new theories called into question nineteenth-century ideas of the universe. After the First World War, this situation became more fraught when scientific institutions were associated with the weaponry and technology that had wreaked unfathomable violence on the battlefields.[27] By analyzing the empirical rhetoric, practices, and technology of paranormal investigators, one can get a sense of the delicate cultural contours of scientific knowledge. In many regards, psychical research vividly reflected the role, influence, and popular understandings of science in the interwar period. More so, its historical formation highlights how specific conceptions of gender, class, and race animated the embodied prerequisites for making scientific claims in the first place.

Situating the Seance: Power, Gender, and Troublesome Bodies

Despite psychical researchers' engagement with scientific knowledge and interwar culture, not to mention the broad popularity of paranormal topics in the 1920s and 1930s, their outlook was not exactly conventional.[28] The level of agency paranormal investigators ascribed to the metaphysical distinguished them from many others in their society, including the majority of natural scientists. Psychical researchers at times rejected, or at least remained skeptical of, the spiritualist assumption that these voices and touches in the seance were actually coming from otherworldly entities. "I am not talking about spirits or spirit manifestations," researcher J. Malcolm Bird explained regarding his approach to psychical investigation. "I am talking about psychic phenomena – phenomena which occur in such fashion as to be a function of the presence of some particular human personality, with its attendant organism, and without explanation in terms of known scientific doctrine."[29]

Regardless of their belief, or lack thereof, investigators and their fellow experimenters named, conversed with, and interacted with alleged discarnate personalities. As T. Glen Hamilton recognized, this approach, however odd at moments, seemed essential for the production of psychical phenomena.[30] Referring to them by name, obeying their specific instructions, and engaging with them verbally and sometimes physically, researchers credited these uncanny personalities with a sense of agency that was otherwise reserved for more materially inclined humans. This study takes such accounts seriously and treats these spirits as the dynamic agents they seemed to be in the seance room.

Those who participated in paranormal investigations approached the seance with several different and complex perspectives. Rather than placing these unsettling subjects within a decipherable box of irrationality and dubiousness, this study aims to understand them in light of what cultural theorist Donna Haraway has referred to as "situated knowledges."[31] As opposed to explaining away such mysterious happenings with an arsenal of deciphering techniques, I instead attempt to grapple with these accounts as significant cultural texts that, like many other transcriptions, contain ambiguities, inconsistencies, and contradictions. Instead of playing the role of rational observer over irrational subjects, I seek to unravel the concept of rationality itself in order to come to terms with the larger cultural significance of interwar psychical research in regard to science, technology, gender, class, race, and the body.

Although the focus is on self-proclaimed "scientific" investigators of the paranormal, I do not make a rigid distinction between "religious" and "secular" perspectives. Psychical research tended to call those categories into question, and the events in the seance room defied boundary-making processes. This study explores women investigators who fluidly moved between empirical ambitions and spiritual convictions. Male investigators, meanwhile, may have presented themselves and their colleagues as pursuing psychical research as an unemotional "cold science," yet they by no means abandoned their religious affiliations. Moreover, heated disagreements over what constituted appropriate controls, or a credible manifestation, revealed that they were not as detached as they might claim.[32]

Rather than asserting some generalized notions of what it meant to be a psychical researcher, I employ a different tactic. As Latour recognizes, social scientists consistently endeavour to find coherence in groups, while too often ignoring their contradictions and volatility. He suggests an alternative, arguing that more "sturdy" and "revealing patterns" can be made evident by recognizing "the links between unstable and shifting frames of reference," as opposed to constantly "trying to keep one frame stable."[33] This practice does not diminish the existence of the group but rests more firmly upon the reiterations of the actors themselves, in contrast to some arbitrary ordering practice that a sociologist or a historian could impose. Applying such an analysis to the rhetoric, ideology, objects, and practice of those who explored psychical phenomena leads to a more involved understanding of what it meant to join hands in a seance circle. It also begins to pry open the very means by which groups and categories are constructed in the first place.[34]

One can readily identify group-making practices among psychical researchers, yet their ideas and practices remained diverse. As a result, this study does not attempt to construct a static and consistent social group out of those who pursued the paranormal. Rather, it follows a set of actors who attempted to forge frameworks of methodology, technological apparatuses, and classifying techniques around and within the paranormal. What emerges is not a unified, impregnable movement but "a moving target" encircled by a vast array of intentions, materials, and performances.[35]

Despite making the actors of psychical research central to my analysis, the task of identifying and acknowledging agency in the seance is complex. This is especially the case when considering the presence of those who seemed to have agency but were not understood by all as *real* in the strict sense of the term – namely the discarnate personalities. Although often said to have bodily likenesses, including hands, heads, and torsos that materialized in the seance room, these intelligences were liminal and indeterminate. Meanwhile, mediums stripped of agency in rhetoric could, at times, wield it rather effectively in practice.

Women predominantly acted as mediums, but some men did as well. Whether working with men or women, however, psychical researchers ascribed stereotypical feminine qualities to each. Much as historian Alex Owen argues, qualities of passivity, compliance, and subjectivity were conceived of as necessary for credible spirit communication to occur and were closely equated with femininity.[36] Characterized by psychical investigators as lacking will, a medium's body – not her mind – transformed into the chief instrument through which the spirits operated. Women mediums were therefore represented as the norm, whereas male mediums, however valued for their gifts, experienced great difficulty in evading classifications of sexual aberrance as they assumed the allegedly submissive role of communing with the spirits.

In such a context of the real and surreal, however, nothing was quite as it seemed. Competing agendas conflicted and personas transfigured as objects materialized, moved, and vanished. Likewise, mediums and their bodies took on identities and forms that were not static, resonating with the claim of theorist Michel Foucault that "the body is the surface of the inscription of events (traced by language and dissolved by ideas)."[37] Rather than submissive blank slates upon which action was imposed, the bodies of mediums, like the ghosts they aroused, proved to be dynamic and sometimes inimical forces, at one moment passively playing the part assigned to them and, at another, subverting any known scripts of behaviour.

Much like scholar Saba Mahmood would caution, the actions of mediums cannot simply be understood in transparent terms of domination and subordination.[38] The assumption that all mediums merely faked trances and phenomena to overthrow patriarchal authority is perhaps a bit too simplistic. Mediums, after all, often willingly cooperated with investigators, at least proposing that they did so for "the interests of truth, without thought of reward of any kind."[39] Yet, even while freely operating within assumed frameworks of passivity and control, mediums and sometimes sitters defined their own sense of agency. Mediums of the interwar era gave over their bodies and minds for use in psychical experiments. Nevertheless, they did so to become a point of communion with another world that provided them a space in which to exercise power in ways that would have been impossible beyond the seance room walls.[40]

Mediums' capacity for agency was frequently limited, however, and viewed as legitimate only if they were understood as under the control of spirits.[41] Only when their minds were entranced by other powers could they then become the submissive bodies necessary for the successful production of paranormal forces. Although psychical researchers tried to distance themselves from spiritualists, their experiments exacerbated the gender dynamics at work within the seance, strengthening the conviction that the medium needed to be absolutely passive.[42] Psychical researchers insisted upon some method of control, being suspicious of any medium who refused to undergo rigid tests.[43] Resistance suggested that the psychic had something to hide. Conscious defiance demonstrated coyness and deception.[44]

Psychical researchers were hardly alone in their convictions, as they buttressed masculine and feminine polarities much like the rest of the scientific establishment. Feminist scholars of technoscience have asserted that scientists insisted upon gendered spectrums of objectivity and subjectivity. The idea of science as a specifically masculine practice emerged from the context of seventeenth-century Europe, which witnessed the construction of "dichotomies between mind and nature, reason and feeling, masculine and feminine."[45] Baconian philosophies of the scientific method were predicated upon the idea that nature needed to be dominated for the furtherance of human civilization. This notion was not necessarily a direct attack upon women, but it used the "well-worn Aristotelian categories" of hot and active masculinity in opposition to cold and passive femininity in order to propose "an active philosophy, one which would act as a formative principle upon a feminine nature."[46] As a consequence of such sexual metaphors, women found themselves alienated from empirical practice, structured as passive and weak subjects, and suitable only for the investigative gaze of the male scientist.[47] Scientific conceptions of feminine weakness and passivity in turn affected a multitude of professions that attempted to don the scientific garb, including medicine and the emerging study of the mind, which was later formalized into modern psychology.[48]

This idea of science was reaffirmed in the seance through a series of discursive and embodied performances. Although psychical researchers represented mediums as passive bodies through which paranormal forces could operate and be experimented with, they also constructed themselves, their methods, and their technologies in such a way as to parallel what they believed to be in line with the rational, scientific principles of their era. Faith in objective, scientific rationality did not die at the end of the nineteenth century. The years following the First World War, which saw for instance the establishment of the Vienna Circle, continued the nineteenth-century legacy whereby scientific knowledge was upheld as "the ultimate achievement of human rationality."[49] Scientists in this equation became "transparent," as Haraway has argued, presenting their ideas and methods as untainted by bodily subjectivity to offer "credibility to their descriptions of other bodies" while "minimiz[ing] critical attention to their own."[50] Psychical researchers endeavoured to take on this distinctly gendered scientific mantle. Employing methods, technologies, and psychological theories to disembody their perception of otherworldly realities, they rendered the mediums under scrutiny as subjective, effeminate, and irrational bodies.

Psychical researchers' reference to the body at this time was not insignificant. The first half of the twentieth century witnessed the rise of several scientific programs that linked biological factors to the embodied formation of sex and gender. Scientists probed how hormones, glands, and molecular genetics influenced physiological characteristics, intent on discovering "sex itself." Historian Sarah S. Richardson notes that during the interwar years "the scientific study of sex underwent rapid consolidation, professionalization and expansion."[51] Yet these investigations were not divorced from their social context. As Angus McLaren describes, scientists frequently joined hands with a host of "eugenically inspired rationalists" who ultimately "sought the disciplining and regulation of the body."[52] Fortifying specific notions of sexual difference, desirable reproductive traits, and distinctive gender roles, scientific discourse of the interwar era bolstered cultural norms by reinscripting opposing formations of masculine and feminine embodiment.[53]

The biological reinforcement of such gendered dichotomies may have been viewed as especially necessary after the First World War. In the 1920s and 1930s, patriarchal structures were perceived as under threat – particularly with women's recent enfranchisement, a growing female labour force, and the rise of the ubiquitous "modern girl."[54] At the same time, an ideal of vigorous masculinity in the late nineteenth and early twentieth centuries seemed to be in peril in the aftermath of the war.[55] In 1914 men marched off to the frontlines, expecting to rejuvenate a militant masculinity that reflected qualities of physical prowess, strength, and courage.[56] The devastations of the war, however, challenged such ambitions and wreaked havoc on men's bodies, as well as their minds. The degree to which the war changed both gender and society more broadly has been debated by historians, as the postwar era arguably witnessed a reassertion of so-called traditional gender roles rather than a reassessment of them. The fears the war experience induced about masculinity and femininity were nevertheless widespread and compelled individuals to respond accordingly.[57]

In a world in which gender seemed to be in the midst of a dangerous flux, psychical researchers reaffirmed notions of passive and weak womanhood through scientific control. Investigators, who were primarily middle- to upper-class, professional men, rigorously managed female bodies, insisting that such control was necessary to establish the scientific credibility of psychical experiments.[58] Conceptions of not only gender but also how gender itself was embodied therefore acted as a powerful organizing principle of interwar paranormal investigations, situating mediums as under the legitimate control of psychical researchers. Male mediums, who potentially threatened the dichotomy of manly authority over womanly mediums, were neutralized by investigators through their biological classification as inherently effeminate.

This power dynamic also manifested itself in other ways as psychical researchers and spiritualists came in contact with worldly and otherworldly figures that fell beneath categories of not only gender but also class and race. Gendered categories of passive feminine embodiment and masculine dominance thus operated within broadly conceived hierarchies of class and race in the seance, highlighting those who were intended to wield authority and those who were not.

The period under investigation fell at the tail end of what some historians have dubbed the "Great Transformation," due to considerable industrial, capital, and urban growth.[59] It was a period of significant working-class labour movements, which especially made their mark on cities where a popular enthusiasm toward the paranormal flourished. Mediums were typically members of the labouring classes themselves, but even when they were not, the spectre of the lower classes remained in various subtle and not so subtle ways. Due to their more fragile economic status, psychics commonly engaged in spirit communication for the purpose of making an income. As they were labourers of a unique sort, their success relied on their ability to be not only convincing but entertaining as well.[60] These two aims went hand in hand, many skeptics remarked, as the more distracted sitters were by floating objects, ghostly voices, touches, or lively spirit guides, the less likely they were to notice any deception on the part of the medium.[61]

As often as psychics were cast as conniving, they were simultaneously fetishized as well, quite possibly to defuse the social challenges that the labour unrest of this era fostered. A pacified, compliant, and feminine version of the working class might have been much more comforting to middle-class sitters than were the riotous union members of their cities. Psychical researchers were by and large from the middle to upper echelons of society, and they compiled the extensive records and photographs of paranormal experiments that now rest in a select number of archives. This study is indebted to such records, yet they too need to be historicized and put into context. These meticulously collected documents emerged from a specific socio-economic perspective and, as a result, expressed a conscious desire to rebrand the pursuit of the paranormal as a more respectable middle- to upper-class enterprise.[62] The constant employment of scientific language, methods, and practice arguably provided such respectability. Moreover, it legitimized investigators' power over others who seemed beyond such control in the world outside the seance room's walls.

Scientific control justified the regulation of gendered, classed, and finally racialized bodies. At least in terms of the network of psychical researchers and mediums that this study addresses, the sitters and even mediums were white. Yet despite such homogeneity, racialized spirits played significant roles in contacting the dead. These spirits needed to be properly controlled and managed according to the expectations of white seance participants. Such management affirmed white privilege and power, with the racialization of the spirits acting as a foil to the supposedly race-neutral, white sitters.[63] At the same time, orientalized ghosts created the necessary conditions for seance participants to rediscover an individualistic metaphysics, while infusing their scientific endeavours with perceptible, and thus empirically measurable, instances of ghostly encounters.

Operating within a number of gender, class, and racial frameworks of power, the seance seemed to be anything but a closed space. Sitters, investigators, mediums, and even spirits hearkened to social structures well beyond the walls of their experimental rooms. Far from a vacuum where individuals pursued eccentric goals irrelevant to the world around them, the seance remained imbued, even haunted, by interwar tensions, anxieties, and ambiguities as investigators endeavoured to fashion a scientific enterprise. The unique space of the seance, however, was more than simply a pale reflection of the interwar world. The struggle for authority between investigator, psychic, and spirit laid bare the deeply political relationship between gender, empiricism, and technology, while also providing the opportunity for embodied reconfigurations.

Sensational Apparitions: Method, Theory, and Organization

By closely examining the scientific identity psychical researchers attempted to adopt for themselves, the mediumistic and ghostly subjects they studied, their records, methods, and technology, and the psychological and scientific classification they employed, this study aims to construct what scholar Steven Connor refers to as a "cultural phenomenology" of these seances. In contrast to Connor, however, I argue that such a perspective does not diminish the gender, class, and race politics of the seance. Rather it is my contention that a meticulous reading brings these struggles to the fore with even greater clarity.[64]

That said, I do not exhaustively illuminate every psychical researcher, medium, or series of experiments conducted at this time, and in certain instances I refer to some rather high-profile cases only in passing or from an atypical perspective. Of all these incidents, the experiments with Mina Crandon, or "Margery," stand as the most famous. Her mediumship became widely known in North America and Europe throughout the 1920s and 1930s, forging how many regarded the paranormal, to the delight of some and the deep chagrin of others.[65] As a result, her mediumship has attracted the attention of a number of scholars. For the most part, however, these narratives have focused almost exclusively on her critics, who deemed Crandon a fraud.[66] Some of these studies, for instance, have examined the damning investigations conducted by famous escape artist Harry Houdini during his notorious involvement with the case.[67] By comparison, very little attention has been paid to Crandon's perspective. As a result, the cynical attitudes of some of her investigators have been largely affirmed rather than analyzed.

By scrutinizing numerous reports, letters, graphs, photographs, publications, and objects produced by psychical researchers, this work aims to reveal the dynamics of seances themselves. Mediums in these contexts were largely silenced throughout experimental records, their stories, attitudes, and experiences being filtered by their examiners. Yet through a more intimate reading of these sources, one can gain a few passing glimpses of individual psychics and their responses to the prods, gropes, and scrutiny of their investigators. Much like several scholars of colonialism, I seek to grasp not only what was recorded but also what was omitted. By identifying absences, gaps, and "unintelligibilities," I aim to detect the subtle, muted voices of those who rested at the centre of these investigations but often had limited power over them.[68]

Pivotal to my reading of these sources is a close examination of physical perception and the embodiment to which it hearkened. Whereas the politics of gender has been closely analyzed in the literature of spiritualism, the role of bodies has received much less attention – a gap that this study endeavours to address. The history of the body stresses that the body is a product of history and is thus not invariable but "always becoming," as historian Lisa Helps describes.[69] The body can therefore not be separated from the world out of which it emerges. It is, as historians Steven Shapin and Christopher Lawrence describe, "culturally embedded and culture-constituting."[70] This conception of the body is indebted to Michel Foucault, who claims that the body is assembled by systems of power that endeavour to structure and manage it.[71] Theorist Judith Butler counters this position, arguing that despite such regulation, the unstable body evades or exceeds categorical discipline, calling into question the "hegemonic force" of sexual and medical norms.[72] Yet an analysis of how the body is culturally and historically situated in no way precludes its physicality, as historian James Opp insists, but rather illustrates "that human experience is mediated prereflexively through the material body."[73] Taken together, the analyses of these scholars suggest that the materiality of the body is an unpredictable product of history and culture. If the body is formed in this way, so too must be its perceptions.

The history of the senses cannot be attributed entirely to interest in the history of the body. This seems especially to be the case in that calls for a history of the senses predate the history of the body, beginning with historian Lucien Febvre in 1941.[74] The history of the body and the history of the senses, however, have not developed in isolation from one another.[75] As anthropologist David Howes and historian Mark Smith have insisted, the perceptions, much like the body, are not universal but need to be historically situated.[76] I therefore consider the body alongside the senses to discern how psychical researchers ordered their own bodies and others in relation to their physical perceptions of the paranormal.[77] The result is an intersensorial analysis that considers how the supernatural materialized in the seance room in relation to material and spiritual bodies.

The following chapters offer a detailed analysis of the ways that psychical researchers constructed a science of the paranormal, examining interwar entanglements between this world and the next through investigators' rhetoric, technology, and practice. Psychical researchers attempted to manage their psychical laboratories according to rigorous standards, honing their methods and technologies to objectively perceive the otherworld. However, their subjects proved difficult to control. Researchers contended that mediums were the perfect bodily machines for producing paranormal effects. As a consequence, these unusually gifted individuals rested at the crux of psychical experiments, virtually dictating when investigations would succeed or fail. Together, investigator, medium, and spirit adopted, engaged with, and remade the gendered meanings ascribed to empiricism and technology through embodied performances that could at once affirm and transgress normative cultural mores of the interwar world.

1
The "Scientific Self": Performative Masculinity in the Psychical Laboratory

> I believe that science, that is, the scientific method, is the only source of evidence in anything.
> – James H. Hyslop, "Signs of the Times"[1]

To shape the pursuit of the paranormal, interwar psychical researchers did much more than introduce the scientific method into the seance room. They first adopted an identity suited to their challenging enterprise. Men such as W.R. Wood, T. Glen Hamilton, Le Roi Crandon, J. Malcolm Bird, Eric J. Dingwall, and Henry Clay McComas endeavoured to represent themselves as credible and rational investigators. Distancing themselves from the subjective influences of the body, male psychical researchers insisted upon their ability to perceive the paranormal from a position of objectivity. Affirming standards of precise measurement, rigorous controls, and experimental consistency, these investigators engaged in what scholar Andrew Pickering calls "a *performative* image of science" by enacting a particular set of attributes believed to be necessary in order to obtain authoritative knowledge of the natural world.[2]

These performances were often embodied, but to reify their gendered prerequisite, they were not recognized as such. Female investigators, as a result, were denied the same level of recognition as their male colleagues. Although women like Lillian Hamilton and Jenny O'Hara Pincock experimented with the spirits alongside men and used the same mediums while upholding a similar set of theories, their femininity posed too great an impediment to rational investigation. Characterized as impressionable and spiritually sensitive, women straddled the alleged divide between scientific study and religious devotion. Their ability to bridge scientific and spiritual desire provided only further reason to deny them the same authority as the men they worked beside. Like "invisible technicians," women such as Hamilton and O'Hara Pincock supported and recorded the findings of psychical scientists and proved vital to creating optimal seance conditions.[3] They were nevertheless refused the status of scientist, as their male counterparts deemed them unable to escape the mental and physical limitations assumed to be inherent to their feminine embodiment.

As an integral part of their scientific practice, psychical researchers created a laboratorial stage that marginalized the stereotypically feminine elements of communing with the spirits. Emptying the space of all its domestic, familial characteristics, psychical scientists restructured the room so that it reflected empirical standards. The seance-turned-laboratory reinforced their claims to scientific authority. Moreover, it provided the backdrop to psychical researchers' elaborate enactment of what historians Lorraine Daston and Peter Galison refer to as a "scientific self."[4] Dedicated to objectivity and rationality, investigators presented themselves as the perfect individuals to conduct a science of spirits. Through such performances, they asserted their authority and thus upheld the same patriarchal hierarchies that existed in a much broader range of scientific enterprises.

Scientific Selves: Objectivity, Manliness, and Psychical Research

An assumption of a rational, masculine identity required investigators to assert binaries between spiritual and secular perspectives by minimizing their own religious identities. United Church minister W.R. Wood of Winnipeg, Manitoba, maintained that he had put aside his faith convictions to pursue the science of psychical research. In an article entitled "Science and Survival," he distanced both himself and psychical research from spiritual passion, insisting he did not wish to speak "from the point of view of religion, but from that of science." Wood then went on to credit the progress of psychical investigations to "men of informed and disciplined minds" who sought to uncover the scientific value in "certain mysterious occurrences which up to that time had been regarded as wholly illusory, the products of imagination or of superstition." In contrast to materialists who argued that science should "be limited to things and occurrences directly associated with matter," Wood argued that these "men of courage ... dared in the interests of a larger and fuller knowledge to push their researches into regions profoundly mysterious and wholly uncharted by previous explorers." Defining psychical researchers in relation to their skilful use of mind rather than body, Wood claimed that his own and other psychical researchers' interest could be trusted to unearth "ascertainable and demonstrable proof" of otherworldly phenomena.[5] Much as historian Janet Oppenheim has identified, psychical experimenters like Wood asserted "their absolute devotion to the standards of open, rational, empirical inquiry set forth by modern science."[6]

Nevertheless, only a certain type of individual could convincingly practise such dedication to the objective study of paranormal happenings – a standard that remained underpinned by specific masculine and upper-middle-class distinction. Psychical researchers were not, Wood insisted, the "credulous and gullible people" most frequently associated with spiritualism but were educated men of a particular social calibre. "Cambridge men of a severely critical and scientific type of mind" founded the British Society for Psychical Research in 1882, Wood pointed out, only to be followed by innumerable pre-eminent professionals of medical, scientific, and academic learning who pursued the study of psychical phenomena. The social position of investigators bolstered their credibility to examine psychical phenomena.[7] By means of such tactics, investigators bound themselves to broader cultural paradigms that linked veracity with a particular social status and gendered embodiment – or more precisely, disembodiment.[8]

At the heart of scientific pursuits of the paranormal lay the concept of objectivity. Daston and Galison argue, "To be objective is to aspire to knowledge that bears no trace of the knower – knowledge unmarked by prejudice or skill, fantasy or judgement, wishing or striving. Objectivity is blind sight, seeing without inference, interpretation or intelligence."[9] Paradoxically, this erasure could be achieved only by nurturing the development of a specific sort of self. As Daston and Galison recognize, men of science exhorted each other to aspire to standards of discipline, denial, and "self-mastery" in pursuit of "the assiduous cultivation of a certain kind of self."[10]

The question remains "what kind of self?" Donna Haraway describes an invisible and "modest" self: "the man – the witness whose accounts mirror reality – must be invisible, that is, an inhabitant of the potent 'unmarked category' ... specifically modern, European, masculine, scientific from the virtue of modesty."[11] Influenced by the broader scientific community, North American psychical researchers of the 1920s and

1930s claimed such an identity for their own. Distancing themselves from embodied, subjective, and often effeminized forms of religious fervour or sentimental emotion, they built an image of themselves that reflected a commitment to rational and impersonal paranormal experimentation. Paralleling Wood's perspective, this identity was by no means neutral but carried with it certain gendered and class connotations. Embracing an idea of the scientist similar to what Haraway refers to as the "*vir modestus,*" psychical researchers represented themselves as being "high status," "disciplined," and able to practise "ethical restraint."[12]

Wood's perspective was undoubtedly shaped by his participation in a series of experiments conducted by medical doctor T. Glen Hamilton of Winnipeg. Reflecting upon his sittings with the group, Wood expressed admiration for their empirical rigour, as he claimed to be "fully convinced of the validity of their experiences."[13] He was hardly alone in his convictions. Perhaps one of the most renowned Canadian psychical experimenters, Hamilton gained an international reputation for his scientific standards. The vast array of graphs, notes, and photographs that he collected throughout the investigations he conducted from 1921 until his death in 1935 were some of the most impressive records of paranormal phenomena in the world.[14] Hamilton had established himself as an expert surgeon and prominent public figure before he began his psychical investigations in 1918.[15] Arguably, he seemed to have many personal reasons for conducting psychical experiments. The various struggles that Winnipeg experienced both during the war and afterward as a result of the influenza epidemic were possible motivations for communing with the spirits.[16] As one of the main general-health practitioners at the Winnipeg hospital, he gained first-hand knowledge of influenza's ability to wreak death and destruction. His knowledge became even more intimate with the death of his young son Arthur from the flu, which scholar Esyllt Jones argues was Hamilton's chief incentive for communing with the spirits.[17] Jones's argument is not without merit. Nevertheless, multiple other factors also seemed to be at work, especially when one considers not only the breadth and rigour of experiments that Hamilton conducted in the years to follow but also his continued insistence that his interest in paranormal phenomena had arisen only when its scientific value became evident. A prominent advocate of the science of medicine and its empirical methods, he represented himself as rational and objective, both professionally and personally.[18]

Although seemingly more than happy to speak with spiritualists, as well as church congregations and clubs, Hamilton differentiated his views from what he perceived as "a purely sentimental and religious standpoint" in both his publications and his private letters.[19] Writing to psychotherapist Walter Franklin Prince, who was also an ordained minister and the chief director of research for the American Society for Psychical Research (ASPR), Hamilton assured him of the validity of the mediumship with which he experimented. Hamilton detailed how he and the other experimenters in his group sheltered both themselves and their relatively uneducated and presumably impressionable mediums from the "admixture of the numerous superstitions and notions which seem to prevail widely among so-called spiritualists."[20] In response to a Nova Scotia woman who asked him to provide details of spiritualist revelations, he replied, "I am not a spiritualist," insisting instead that he viewed the study of psychical phenomena "as a cold science."[21]

Hamilton's reluctance to admit to subjective elements in his investigation of the paranormal did not necessarily negate all religious perspective. Rather, he endorsed a certain brand of faith commonly referred to by scholars as "rational religion." Inspired by eighteenth- and nineteenth-century ideals of positivism and reason, this religious orientation melded scientific and spiritual understandings of the world.[22] Continuing well into the twentieth century, it reflected a liberal theology that embraced a rationalization and ordering of faith experiences through the language of science, psychology, and self-fulfilment.[23] Notably, twentieth-century incarnations of this distinct brand of religion, such as the Oxford Group, typically catered to educated middle- to upper-class sensibilities of respectability and intellect, making "religion and modernity seem compatible."[24]

Advocating the values of reason and sombre self-control, Hamilton disdained expressions of fundamentalist Christianity that he perceived as "fanciful," "misleading," and most importantly, "unscientific."[25] Writing to a concerned woman, Hamilton tersely placed a book she had mailed him within a broader context, identifying how it demonstrated what he viewed as one prevalent pitfall of this brand of Christian faith: "You will notice that it is published in Tennessee. It is the sort of thing that one finds very prevalent in the Southern States where the anthropomorphic idea of God is so prevalent." Promoting "the interests of that higher type of Christianity which it is our privilege to enjoy," he paternalistically urged that his female correspondent "not allow teachings of this kind to get a strong grip on you."[26]

Hamilton's dedication to a rationally ordered spirituality remained central to how he forged his perspective and identity as a credible observer of psychical phenomena. Although recognized as a man of "deep religious convictions," he also felt the need to insist upon his qualifications as an objective scientist.[27] Hamilton repeatedly voiced the conviction "that in no sense is psychical research a religion." Although recognizing its spiritual value, Hamilton argued that paranormal investigation "has equally a message for science – many facts to physiology, biology, psychology, physics and other departments of inquiry must eventually prove of great value." In light of this interconnectedness between psychical and natural sciences, Hamilton ascertained that the field of "metaphysics is not a religion but a patient, persistent and on the whole, scientific inquiry into the actuality, nature and implications of those obscure phenomena."[28]

Ann Taves, a professor of religion, identifies how psychical researchers' efforts to maintain a balance between science and faith became ever more difficult in the early twentieth century. The emerging fields of psychology and psychoanalysis took pains to pathologize religious fervour, casting it as the polar opposite of rationalist ideals and irreconcilable to scientific pursuits.[29] Well aware of this precariousness, Hamilton regarded his psychical experiments as "a scientific subject so difficult," especially when it came to imparting the content of such studies to public audiences. He consequently remained "impressed with the necessity of slowly but surely establishing and maintaining public confidence in scientific work and careful judgement."[30] Only very cautiously did he make his experiments public, and he refrained from publishing his results until the late 1920s. Hamilton also continued to be timid about making any definite conclusions about the otherworld, representing both himself and those who joined him in the experiments as driven by the requirements and rigours of the "scientific method" rather than by "motives of sentiment ... or religious beliefs derived from any sect whatsoever."[31] By reifying a binary between religion and secular reason, Hamilton aligned himself with what historian John Lardas Modern refers to as "natural neutrality."[32] Consistently estranging emotional spiritual conviction from his work, Hamilton structured his psychical experiments, his religious sensibilities, and himself as unerringly rational.

Those who joined Hamilton in the seance circle similarly endorsed this sort of scientism and rigid objectivity. A number of women participated in Hamilton's experiments and frequently proved vital to seance procedures. Yet, when listing the members of his experimental group in an article for the British journal *Psychic Science,* Hamilton commended the group as comprised of "good men and true."[33] Rather than being a singular mistaken omission, this lack of recognition of the women at work in his psychical laboratory was part of a pattern evident throughout his records and publications. Hamilton emphasized the presence and participation of men in his psychical experiments, reinforcing the conviction that masculinity and scientific credibility remained intricately intertwined.

Although Hamilton had previously shown interest in telepathy and paranormal phenomena, from 1921 onward his intermittent fascination was transformed into a lifelong endeavour. In this year, he began conducting experiments with Elizabeth Poole, or "Elizabeth M," as he referred to her in his records – a woman who had worked alongside him as a nurse for over twenty years and who exhibited faculties of an extraordinary kind.[34] In both published and unpublished accounts of these experiments, Hamilton repeatedly noted the other male professionals he invited to join him in his investigations of this very special medium. The predominantly male personnel of Hamilton's experimental room included his brother, medical doctor J.A. Hamilton, electrical expert H.A. Reed, Rev. D.N. McLaughlin, engineer H. Shand, solicitor Harry A.V. Green, businessman W.B. Cooper, and medical scientist and doctor Bruce Chown, among others. Hamilton credited these professionals with helping to maintain the scientific and thus fraud-proof conditions. When he did mention the presence of women, he did so only in passing – making their participation, at least rhetorically, secondary to the success and legitimacy of each experiment.[35]

Hamilton and his fellow male colleagues conducted several collaborative experiments that attempted to rationally discern inexplicable happenings. They observed the seance table vibrate, levitate, or be pressed to the floor by a mysterious force once the table had first been in contact with Elizabeth Poole. This display of telekinetic power convinced Hamilton and his colleagues of the need to design a series of different experiments. Many of these tests, which Hamilton photographed, involved male investigators attempting to counter the direction of the force. Even if women were present in the room, their existence beyond the frame of the camera's lens excluded them visually, if not actually, from the experimental procedures (see Figure 1).

The procedure of the tests was fairly straightforward. When the table levitated, Hamilton and other men pushed the table down to try and judge the strength of the force. Similarly, when the same said unidentified energy depressed the table, the experimenters attempted to lift the table, at times exerting a force they measured to be "40-60 pounds."[36] Throughout these experiments, the male investigators characterized their efforts as little more than an external, mechanical force comparable to a weighing scale – a device they would use in later experiments, as discussed

FIGURE 1 "Tilting Table," July 1923. A photograph of one of the levitating table experiments with the medium Elizabeth Poole in the Hamiltons' psychical laboratory in Winnipeg, Manitoba. D.B. McDonald, an occasional co-experimenter, is holding the other side of the table in order to approximate the strength of the paranormal force.

in Chapter 4. The investigators remained untouched by the paranormal energies and knew when to stop only the moment that the medium Elizabeth Poole "cried out with pain" due to a "wrenching" sensation in her abdomen as the investigators manipulated the table.[37] Unlike the medium, the investigators maintained a position of distant neutrality, their bodies unremarked upon, whereas Poole's body remained entangled with the phenomena. With the agency of their bodies diminished, Hamilton and the other investigators assumed the mantle of objective and authoritative scientists acting upon preternatural forces embodied by the medium.

Experimental Networks: Joint Sittings and Manly Collaboration

Hamilton carefully organized a group of fellow experimenters who were masculine, objective, and untainted by subjective embodiment so that they might confirm the legitimacy of his pursuits. Yet Hamilton extended his circle even wider as he aligned his methods, applied theories, and results with a much broader scientific community well beyond the walls of his psychical laboratory. This community acted together to construct a science of ghosts built upon similar experimental procedures, theories, and repeatable phenomena. He compared his experiments and results with those of psychical investigators such as Albert von Schrenck-Notzing, William Crookes, William Crawford, Le Roi Crandon, and others to affirm his own integrity, as well as the validity of the phenomena observed and documented. "In short where manifestations of a physical nature are both genuine and powerful, invariably we find the same leading characteristics," Hamilton asserted in his notes, "whether the phenomena appear in England or in France, in Ireland or in Canada, and whether the medium be D.D. Home, Eva C., Kathleen Goligher, or Mrs. Poole." This observation, in turn, led him to the general conclusion that "nature alone is the invariable creator of these rare and little known phenomena."[38] Hamilton insisted that he was by no means isolated in conducting and recording paranormal experiments but was contributing to a broader base of knowledge upheld by a multitude.

Hamilton made these links tangible by conducting joint sittings with other well-known psychical researchers. Such cooperative, transnational investigations helped to endorse his own and others' experiments as a strictly scientific, objective, and fundamentally masculine enterprise. The most famous and influential of these experiments was with psychical researcher Le Roi Crandon and his mediumistic wife, Mina Crandon (née Stinson).[39]

A prestigious Boston surgeon who had taught at Harvard Medical School for several years, Crandon was known for his "hard-boiled materialist creed."[40] His convictions did not waver despite his close encounters with death, particularly while working as a surgeon with the naval hospital that operated out of Chelsea, Massachusetts, during the First World War. Yet in 1923, upon reading the accounts of psychical researcher William Crawford, he began to take claims of paranormal activity more seriously. After a convincing encounter with a professional psychic during the same year, Le Roi and Mina Crandon, along with a select number of close associates, began to experiment with table-tilting phenomena at their Boston home, located at 10 Lime Street. This early experimental group was comprised of medical doctor Edison W. Brown; his wife, Katherine Caldwell Brown; and her brother, dentist Frederick Caldwell. Participants also included Frederick Adler, agent-superintendant of Crandon's office building, along with British army veteran Alexander W. Cross. Together, they formed what became known as the "ABC group" and would first discover the unique capabilities of Mina Crandon. Systemized knocks evolved into animate furniture, floating objects, enigmatic writings, and mysterious whispers from charismatic personalities – the most significant claiming to be Walter Stinson, Mina Crandon's long-dead brother.[41]

As the months passed, their circle expanded to include Boston physician Mark Richardson, renowned for his work on typhoid vaccination and, later, infantile paralysis. Richardson's wife, as well as a handful of relatives and friends, also occasionally attended. The gravity of these experiments heightened as the mediumship became more complex and extraordinary. As with Hamilton's experiments, women were permitted into the seance, but men were typically granted much of the authority to decipher the authenticity of the mediumship and its attendant effects. Crandon allowed other participants into the circle to make more formal inquiries into the mediumship. These new participants were by and large "highly regarded" men whose presence and favourable testimony further legitimized the phenomena.[42] Their involvement, however, eventually led to Mina Crandon being thrust into the limelight.

By early June 1923, Harvard psychology professor Abraham A. Roback had joined the circle. Later that year, he brought with him other Harvard colleagues, including British psychologist William McDougall, who had been recruited by William James to join the faculty of the Ivy League school three years before.[43] Together, they established one of the first groups to undertake a formal examination of the case. Although the group quickly dissolved after one of its members claimed he had found evidence of fraud, others came after them to similarly insist upon an independent and formal scientific investigation. In December 1923 the Crandons left for Europe, where two famous psychical researchers, physician Gustave Geley and physiologist Charles Richet, examined Mina Crandon in Paris, followed by an official investigation at the British College of Psychic Science in London and finally an evening of experimentation at the British Society for Psychical Research.[44]

While at the Society of Psychical Research, the Crandons first met Eric J. Dingwall, its research officer. Dingwall was initially trained as an anthropologist, but an intensifying interest in the paranormal led him to carry out extensive experiments with mediums across Europe and North America for several decades. Throughout his long career as a psychical researcher, Dingwall insisted upon his own objectivity and commitment to science. In a manual he published just a year before meeting the Crandons, Dingwall presented himself as an impartial professional who insisted upon the importance of gathering "evidence based on facts, and facts only," while exhorting others to be like himself and "preserve the scientific attitude."[45] Largely recognized as a skeptical inquirer, Dingwall was nevertheless impressed with the phenomena that emerged from Mina Crandon. Referring to the levitation of a table in a brightly lit room during his sitting with her as "striking," Dingwall admitted that it aroused his interest to such a degree that he resolved to follow the case further.[46]

Despite these transnational encounters, the results of the Crandon experiments remained unpublished and relatively unknown until J. Malcolm Bird, associate editor of the popular publication *Scientific American,* also heard about the mediumship. Bird instigated the magazine's expressed interest in psychical research from the early 1920s onward. He joined the Boston circle in November 1923, shortly after the Harvard investigation had ended. Bird gradually became fully accepted by the group and gained extraordinary access to Crandon's clairvoyant wife for many years. He was also responsible for much of the publicity that the case received.

Although Bird endeavoured to attract publishable articles pertaining to credible investigations, he found the content of the few that he received unconvincing. Consequently, with the support of his managing editor, A.C. Lescarboura, Bird organized a committee sponsored by the publication that would examine a number of psychics, awarding anyone proven genuine an amount of $2,500. This committee included William McDougall of the now-defunct group of Harvard investigators; Daniel Frost Comstock, physicist and retired member of the faculty at the Massachusetts Institute of Technology; Walter Franklin Prince, minister, psychotherapist, and the research officer of the ASPR; Hereward

Carrington, widely acclaimed psychical researcher; and Harry Houdini, world-renowned illusionist and escape artist. After the *Scientific American* committee failed to attract any adequate mediums, Le Roi Crandon finally expressed his wife's willingness to participate, but there were two conditions. First, they were not to receive any award so as to avoid claims of profiteering on their part. Second, Mina Crandon's identity was to be concealed by giving her the pseudonym "Margery" throughout all the public reports.[47]

The following series of investigations eventually made the case internationally known, especially Houdini's notorious battles with Mina Crandon, or "Margery," as she was then more widely known, which scholars such as John F. Kasson and Marina Warner have documented.[48] This investigation would be followed by more in the coming years, transforming the now Margery into a public spectacle on whom psychical researchers, the press, and numerous commentators reported from the 1920s to her death in 1941.[49] As a result, Le Roi Crandon was launched onto the world stage of paranormal investigations, seeming to be, as one investigator described him, the exact sort of person whom "everyone directly or remotely interested in psychic phenomena had long desired." Crandon was a man who "could be counted upon to take and maintain a really scientific attitude."[50] He began contributing to the journal of the ASPR, debating with other psychical scientists to establish himself as a prominent, albeit controversial, expert in psychical science.[51]

Crandon represented himself as striving for the same standards of objectivity and rationality as others in the burgeoning field. Much like Hamilton and Dingwall, Crandon insisted on strict scientific methods throughout the majority of the psychical experiments he conducted, maintaining that the controls administered throughout the seance should be "so rigid as to eliminate the question [of fraud] from all discussion."[52] Even those who questioned the veracity of Crandon's experiments with his wife applauded him for his gifts as a scientist. An acclaimed medical doctor and surgeon firmly dedicated to the "interests of scientific truth," Crandon presented himself to even his critics as a person with many qualities necessary to conduct credible and worthwhile experiments, including "manual dexterity," "tenacity of purpose," and a "fervent rationalism."[53]

Hamilton became familiar with Crandon's Boston experiments by reading various published articles surrounding the Margery mediumship, especially after reigniting a previous acquaintance with Crandon in 1925.[54] Hamilton received his first direct introduction to what became known as the "Margery phenomenon" in 1925 when he visited the Crandons at their Lime Street home in Boston. His next involvement with the Margery case occurred at his own psychical laboratory the following year. Hamilton and Crandon applied similar methods of investigation, including stripping, inspecting, redressing, and restraining the medium during the seance. Although some of the trappings and apparatuses used by the two psychical researchers differed slightly, many were the same, including the use of a standard seance cabinet, a bell-box, and a red light.[55] When Hamilton was faced with an unfamiliar piece of equipment, Crandon and his co-experimenter, Mark Richardson, made sure to acquaint him with its workings. As Hamilton reported of one instance, "I was privileged to take part in a tête-à-tête test with Dr. Richardson's justly famous voice-cut-out machine and found it to be absolutely fraud-proof and 100 per cent effective."[56]

After this experience, Hamilton repeatedly expressed support for Crandon's experiments, aligning them with his own psychical investigations in both method and import. Less than a year after his experiments with Le Roi and Mina Crandon in his Winnipeg laboratory, Hamilton wrote to Bird, "I am prepared to state most emphatically that the 'Margery' phenomena is absolutely genuine."[57] In this situation and others, Hamilton reiterated how he and those with whom he joined hands in the seance uniformly conducted objective experiments "of a purely scientific nature."[58]

Hamilton and Crandon were far from alone in identifying themselves in this way but joined a chorus of others who viewed the paranormal as one more phenomenon that required rational and methodical investigation. Adopting the rhetoric and practice of other physical sciences, these investigators reiterated again and again their own attitude toward the study of the spirits. Although often rejecting a strictly materialist perception of the universe, psychical researchers insisted that their outlook was different from that of allegedly sentimental spiritualists. Dingwall wished that more "people would be content in employing scientific methods" like himself, in contrast to "investigat[ing] in a hazy and personal way which has practically no scientific value at all."[59] Bird claimed that spiritualist conclusions, drawn "out of faith and emotion," were "too much for most of us." In contrast, he remained interested in such phenomena "scientifically," being determined to "escape from superstition" by verifying the occurrence of psychic powers before trying to supply a metaphysical explanation for them.[60]

Henry Clay McComas was a psychologist trained at Princeton University, as well as a noted psychical researcher associated with the ASPR.

In this capacity, he also investigated Margery, in addition to a number of other high-profile mediums over the course of a long and varied career.[61] McComas explained that although he admired the very fruitful seance conditions that a set of "believers" could produce, he remained doubtful of their powers of perception in the challenging environment of the seance. McComas conceived of such "uncritical and sympathetic people" as necessarily women, whose presence might lead to marvellous phenomena but who lacked the deductive and observatory faculties of "critical men."[62] Claiming "there is nothing in the séance to help the naive," he lamented that a person of this inclination has "no training to cope with the situation." As a consequence, any phenomena that the individual "sees, hears and describes are just about worthless."[63] Juxtaposing himself against this vulnerable, superstitious mindset, McComas emphasized his scientific authority and laboratory experience. His training as a psychologist made him versatile, rational, and attuned to different manifestations of "consciousness." Such expertise made him "just the man to tell us whether there is anything resembling intelligence in the curious phenomena that goes [sic] under the heading of Psychic Research."[64]

Haraway describes the "modest witness" as one who stands as the "authorized ventriloquist for the object world, adding nothing from his mere opinions, from his biasing embodiment ... His subjectivity is his objectivity."[65] Psychical researchers embraced and reflected such an identity, taking on what they believed to be the qualities necessary to conduct verifiable and scientific investigations of supernormal phenomena. Rational, objective, and masculine, psychical researchers such as Wood, Hamilton, Crandon, Dingwall, Bird, and McComas presented themselves as able to exercise the experimental rigour that a science of ghosts required. Yet, much as McComas and other psychical researchers asserted, this social identity entailed a series of exclusions that restricted those who could not reach the standard of masculine rationality.

Sean-tific Femininity: Women and the Gendered Boundaries of Psychical Science

In describing the masculine "*vir modestus,*" Haraway details the rhetorical and practical exclusion of women from the creation of knowledge. "Within the conventions of modest truth telling," she asserts, "women might watch a demonstration; they could not witness it."[66] Much as historian Ruth Watts identifies, this trend of not including women in the production of scientific knowledge continued well into the 1920s and 1930s. Despite enormous gains by feminists in opening up scientific and medical professions to women from the late nineteenth century onward, femininity continued to be associated with intellectual ineptitude. Women, still defined largely by their bodies rather than by their minds, were assumed to be incapable of the rationalism necessary for scientific pursuits.[67]

In attempting to construct and portray psychical investigations as a science, researchers enacted boundary-making practices that fell along gendered lines. The scientific self envisioned as explicitly male became the standard of a credible investigator. Nevertheless, numerous women attempted to conduct psychical investigations in apparent contradiction to the masculine image of the scientific identity. Yet these women did so only within certain bounds and commonly evoked not their own authority to observe and collect reliable data but the expertise of scientific men.

Lillian Hamilton, T. Glen Hamilton's wife, went to great lengths to support her husband's research and became instrumental in compelling him to continue his investigations of the paranormal. It was she who began to hold sittings with the medium Elizabeth Poole – the woman who would act as the main medium for the Winnipeg experiments for several years following. Although T. Glen had begun experiments with telepathy in 1918, he had decided to give it up, partly due to professional commitments but also for fear that psychical research may be "a dangerous business" that he would be best to avoid.[68] Thus, even when prodded, he resisted dabbling in such practices for a time. Lillian, in contrast, was not to be dissuaded once her interest was "aroused in the possibilities" of such experiments.[69]

Lillian Hamilton continued holding sittings, determined to discover something of value, and in July 1921 her expectations were met. At this time, she observed that when Poole placed her hands on the seance table, it began to tilt on two legs, the other side levitating and offering due resistance against efforts to push it down. "What was holding it in place?" Lillian Hamilton asked rhetorically. "Having recently read Dr. Crawford's account of the telekinetic phenomena which he had obtained with Miss Goligher in Belfast, I jumped to the conclusion that in Elizabeth we had perhaps discovered a medium with potential of the same type."[70] Lillian Hamilton recorded this significant seance on July 24, 1921, in short staccato phrases reminiscent of a scientific report: "Mrs. Poole and Lillian H. present. They place their hands on the table. In a few moments the 'power' is exceedingly strong – the table tilts on two legs ... L.H. tried to depress it back to the floor but found the table seemed to be resting on a sort of 'air cushion.'"[71]

Despite the promising experiment, Lillian Hamilton met with difficulties when she attempted to pursue and present these psychical experiments to others. Her struggles with being recognized as a scientific experimenter came first of all from her husband. T. Glen Hamilton did not initially believe the account of his wife or the medium. Drawing attention to his wife's body, which he assumed could easily lead her feminine mind astray, he suggested that "probably the 'force' was due to unconscious muscular activity."[72] He remained unconvinced until he experienced first-hand such extraordinary powers. On July 31, 1921, T. Glen relented to Lillian's urgings, and they held a seance in which table tilting again occurred. T. Glen was unable to push the table back down in place due to the incredible force. Perhaps unsurprising, the recorder of the seance, Lillian, reported that she "was very much amused to hear TGH 'grunting' ... as he struggled with the table to push it down."[73]

After this experiment, T. Glen Hamilton finally viewed Poole's mediumistic gifts as potentially valuable. Tellingly, only once he himself had experienced the phenomenon was he "convinced for the first time of the reality of psychic force," which he thus deemed worthy of his scientific investigation.[74] A drive to "know the facts of psychical manifestation for one's self" undoubtedly reflected the individualist ideals of his interwar context.[75] Yet it also made Lillian Hamilton merely incidental to his own interests in the paranormal.

Feminist thinker and physicist Evelyn Fox Keller identifies a powerful mythology embedded within the modern scientific enterprise that has cast women as "the guarantors and protectors of the personal, the emotional, the particular." Meanwhile, "science – the province par excellence of the impersonal, the rational and the general – has been the preserve of men."[76] In the process of adopting a scientific framework through which to investigate the paranormal, psychical researchers espoused a similarly gendered division of knowledge. T. Glen Hamilton did value the work of his wife and other female participants in his psychical experiments and, by all appearances, enjoyed conversing with women and communicating to them the methods of his research and results. Yet despite his favourable opinion of women, he did not align them with critical reasoning and experimental methods. He insisted, rather, that women's "psychology is different" from that of men. According to Hamilton, women, unlike men, were innately trusting, emotional, uncritical, and "more disposed to accept with less demand for fundamental detail."[77]

Women, in some instances, seemed to adopt this characterization. As much as Lillian Hamilton articulated a sincere dedication to the scientific method, she also conveyed a much more subjective side to the investigations than did her husband. Whereas T. Glen Hamilton insisted upon his objectivity and unsentimental approach, Lillian freely admitted that the psychical investigations of both herself and her husband had led her to the conviction that "the problem was settled: religious faith in survival no longer walked alone." Expressing her desire for faith alongside her scientific persuasions, she embraced the close links between women and spirituality that several historians have identified.[78] According to Lillian, her spiritual belief "went hand in hand with evidence of a scientific nature."[79] This revelation comforted Lillian, who immediately viewed it in light of the loss of her young son Arthur only a few years before: "A new world had opened up – a world of belief that helped me part with Arthur without tears and with an inner joy that one of my beloved at least was safely over and ready for other-world evolutionary endeavours."[80] The connection she drew between these investigations and her dead son was far from unfounded, as apparently the ghost of their child made frequent appearances in Poole's visions, through which she described him "as increasing in age and stature."[81]

Lillian Hamilton adhered to scientific empiricism much like her husband, but she did not equate her dedication to empiricism with an inability to express grief and hope. T. Glen Hamilton did not admit to such emotion, at least in public, quite possibly out of fear that he would lose legitimacy as an appropriately manly and rational investigator. Seen in this light, the ideal of masculine scientific authority constrained him as well as Lillian. T. Glen may have felt prohibited from communicating his grief over the loss of his son or the hope that his experiments provided. Lillian's manner of forging her identity and perspective provided a degree of flexibility and dynamism that her husband could not afford. Nevertheless, it also safely placed her within the confines of ideal domesticity, motherhood, and respectable, middle-class femininity.

Whereas T. Glen Hamilton experienced significant recognition as a scientific investigator, Lillian Hamilton found herself positioned as irrevocably tied to supposed qualities of womanhood, such as impressionability, irrationality, and emotionalism. She consequently remained unable to fully assume the position of a credible investigator in her own right, despite her invaluable service as a researcher, recorder, witness, and experimenter in the Winnipeg seances.[82] Unlike her husband, Lillian remained defined by her "naturally subjective" knowledge and embodiment. Much as feminist theorist Lorraine Code argues, she could therefore never attain the status of "a knower in the fullest sense of the term."[83]

Lillian continued to experiment for years after the death of her husband. Yet, even when she acted independently, her investigations never received the recognition that his experiments had. Her role, at best, paralleled what Steven Shapin refers to as an "invisible technician" – an essential yet virtually unrecognized agent in the context of the psychical laboratory.[84]

Writer and musician Jenny O'Hara Pincock arguably played a more central and visible role in a series of seances with the medium William Cartheuser from the late 1920s to the mid-1930s. Yet much like Lillian Hamilton, she coupled her spiritual and personal desires with her scientific motivations, consistently demonstrating knowledge of the methods and theories of psychical research while retelling her intimate experiences of paranormal encounters. Her interest in communing with the spirits began in the late 1920s when she and her husband, Newton Pincock, an osteopathic practitioner from St. Catharines, Ontario, began searching for alternate methods of curing a mysterious illness that plagued him. On advice from a friend, they travelled to the New York spiritualist retreat of Lily Dale, where they attended a seance with Cartheuser in September 1927.

Cartheuser had spent the previous fourteen years developing his gifts, over which time he had became known for his ability to cajole spirits into speaking through an aluminum trumpet to his earthly sitters. He had been investigated several times by North American psychical researchers and had even participated in a joint sitting with the famous Margery in Crandon's psychical laboratory under the supervision of J. Malcolm Bird only a year before his meeting with the Pincocks.[85] Cartheuser's main spirit control at the time, referred to as "Dr. Anderson," had gained a reputation for his healing abilities, diagnosing and prescribing various types of cures during the seance.[86] The Pincocks described the seance as "better than anything we had ever read about."[87] Although sitting with Cartheuser did not cure Newton, who died the following May, this experience inspired Jenny O'Hara Pincock to probe the supernatural further. She not only continued to visit Lily Dale but also convinced Cartheuser to become the medium for a series of seances she began to hold at her Ontario home.[88]

In 1930 O'Hara Pincock presented her experiences with the spirits that emerged from Cartheuser in her book *Trails of Truth*. Much like Lillian Hamilton, O'Hara Pincock did not refrain from including some of her more subjective reasons for arriving at spiritualist conclusions, including the deaths of her infant children, Jane and Bobby, as well as her husband, Newton.[89] She nevertheless described these intimate experiences as a "personal explanation" that framed, but did not diminish, the empirical relevance of the paranormal phenomena she described throughout the book. A categorized chart of evidence and a lengthy discussion of scientific theories of the paranormal made up most of the introductory chapters, in conjunction with quotations from British physicist Oliver Lodge.[90] Contextualizing paranormal pursuits within what she termed "the Era of the Invisible Force," O'Hara Pincock argued,

> Man has left the kindergarten of invisibility to emerge into the school of electricity and ether. Radium and the radio now fail to excite his wonder. He has become accustomed to dealing with invisible forces and, providing some tangible medium of matter is still left him to handle and to hold, to control or to create, he doubts not their reality.[91]

O'Hara Pincock situated the pursuits of both herself and those who sat with her within a scientific framework, even while clinging to the spiritual significance of such encounters. Her establishment of the Church of Divine Revelation and the Radiant Healing Centre in St. Catharines the same year that her book was published made her religious motivations apparent. This spiritual focus did not mean she abandoned the empirical impetus for conducting these seances. O'Hara Pincock continued to record and compile persuasive evidence of supernormal realities, which in her perspective went hand in hand: "The revelations of spirit return must grow upon the infallible foundation of evidence. Evidence dispels falsehoods and reveals eternal verities."[92]

Others supported her in this claim by recognizing phenomena produced at the seances of the Church of Divine Revelation for their empirical value and "evidential matter."[93] Even the research officer from the National Laboratory of Psychical Research in London, Harry Price, viewed the records from O'Hara Pincock's seances as "authoritative."[94] Perhaps the most significant affirmation of the scientific aspect of her seances with the Church of Divine Revelation was an endorsement by internationally acclaimed Italian psychical researcher Ernesto Bozzano, a man known as "the highest authority on metaphysics."[95] Referring to O'Hara Pincock and the members of the church's circle as "experimenters" and the seances themselves as "experiments," he perceived these seances as full of "veridical information" and thus in line with what he understood as "scientific methods."[96]

Despite such recognition, O'Hara Pincock notably did not impart her own findings in her book *Trails of Truth* but presented the theories of authorities such as Oliver Lodge, James H. Hyslop, William Crookes, Camille Flammarion, and others.[97] When she did describe her experiences with William Cartheuser, they were largely verbatim accounts from the seances themselves rather than her own theories and experimental strategies. As a result, O'Hara Pincock was discursively thrust to the background, regardless of how instrumental she was to the seance itself. Listing various pieces of evidence according to certain scientific categories, she adopted the role of passive recorder rather than a credible investigator in her own right.[98] Much as Alison Winter has suggested of other women who popularized scientific subjects, O'Hara Pincock's role in the seances reflected the common gendered assumption that women remained "suited only to mechanical forms of intellectual activity" and thus could act only as a labourer, "repackaging and disseminating knowledge created by men."[99]

As the experiences of Lillian Hamilton and Jenny O'Hara Pincock demonstrate, the masculine psychical investigator retained the authority to conduct experiments in a way women could not. Psychical researchers undoubtedly recognized women's ability to create formidable seance conditions, even going so far as to claim that the "men in the group are not as essential as the women."[100] Nevertheless, the power of women's presence lay in the ability of their bodies to channel psychic forces, not to conduct scientific investigations. At best, women such as Hamilton and O'Hara Pincock could act as invisible technicians or automatic transcribers of scientific theories and experimental methods. Paralleling the claims of feminist scholars Mary Jacobus, Evelyn Fox Keller, and Sally Shuttleworth, women's bodies rather than their minds operated "as the imaginary site where meaning (or life) is generated" but could "never be meaning makers in their own right."[101]

In fact, under the authority of psychical researchers, the space of the seance transformed from one in which women and their intuitive psychic powers reigned to one in which the feminine was to be controlled, inspected, regulated, and measured. For this level of control to be made possible, however, the entire space itself needed to undergo a radical shift from seance to psychical laboratory. This physical transformation only further contributed to the construction of the masculine psychical scientist. In addition to enacting rationality, self-control, and scientific precision, manly investigators needed to construct the appropriate spaces in which their performance of the scientific self could be further legitimized.

Masculine Realms: The Making of Psychical Laboratories

Psychical researchers practised their science in different types of spaces and locations. Yet between the world wars they placed increasing emphasis upon constructing laboratories. Refashioning seance rooms into austere laboratory spaces of masculine objectivity was, they believed, a necessary step toward transforming their unique branch of study into a bona fide science. "A well-equipped and endowed psychic laboratory," Hereward Carrington claimed, "is, I believe, one of the prime needs of our time."[102] According to Bruno Latour, the laboratory, with all its "material elements," remained the prime site for not merely the pursuit but also "the production of facts." Within its walls, "objects of study" were made legible for the assembly of scientific knowledge.[103] Psychical investigators thus sought to create the appropriate space within which mysterious, even irrational, phenomena could be remade into empirical truths of nature.

The seance room seemed to be an enigmatic place, marked by darkness and inexplicable touches, smells, lights, and sounds. In the hands of psychical researchers, however, it became increasingly bare and utilitarian, a space where men of status scrutinized, measured, and studied both mediums and ghostly phenomena according to scientific methods. Ridding seance spaces of memorabilia, plush furnishings, and domestic objects that might further obscure the origins of psychic powers, psychical scientists aimed to force the immaterial into material realms. Minimalist in its aesthetic, the psychical experimental room revolved around an ethic of impersonalized and objective rigour. The purpose of the space was to enable the identification, analysis, and quantification of the paranormal in order to place it on par with the material phenomena of scientific investigation. The psychical laboratory thus came to reflect Haraway's description of other scientific spaces: an open yet enclosed "theatre of persuasion."[104]

Such a transformation, however, did not come easily; it required that investigators drastically change experimental techniques, procedures, and the space itself. Researchers frequently insisted on moving the medium to spaces deemed suitable for scientific endeavours and more manageable according to their own standards of objectivity and observation. Investigator Henry Clay McComas did venture to the spiritualist retreat of Lily Dale to seek out mediums, a place that by no means met objective, scientific standards. Founded in the late nineteenth century, Lily Dale was intended by its administrators to be a spiritual and natural respite from "the turmoil of industry and commerce."[105] This place was established in a period when the "wilderness holiday" boomed in popularity in North America, specifically catering to an expanding urban middle class with greater disposable income and available leisure time.[106] Equipped with amenities such as electricity and a postal service, in addition to hotels, shops, and cafeterias, Lily Dale fit the image of the modern natural resort.

Its main attraction was its famous spiritualist teachers and array of mediums who offered spiritual manifestations at a reasonable price. Meanwhile, landmarks such as its "Forest Temple" and "Inspiration Stump" added to its appeal as a natural place where seekers of "higher spiritual truths" could visit for economical fees.[107]

"From all over the country," mediums congregated at Lily Dale, McComas claimed. Each placing a sign in the window along "cottaged streets," they promised spiritual uplift, renewed intimacy with dead loved ones, and entertainment, all for only "a few dollars."[108] Typical of mediums elsewhere, most of those who practised their gifts of mediumship were working-class, offering their very unique wares to a middle-class clientele. The space seemed ill-suited for scientific pursuits, resembling a commercial venue in which spirits manifested primarily for profit – a quality that did not sit well with many: "I think it is a pity," Amelia Holden of Toronto maintained, "that Spiritualism is so commercialized at Lily Dale."[109]

Finding New York to be "so very poorly supplied with good mediums," McComas went to Lily Dale in the summer of 1926 to identify a suitable subject for investigation. Although amused by most of the mediums he met, he did find one whom he believed to be "very remarkable": William Cartheuser. McComas acclaimed him to be "among the greatest of group mediums that I met."[110] Such praise, however, came only once Cartheuser had agreed to leave Lily Dale and hold a seance under McComas's direct supervision. Due to its uneasy closeness to what many perceived as frivolous, irrational, and commercialized spirit contact, Lily Dale proved to be too unstable to authenticate mediumship. Such a test could be commenced only in a room that had been inspected and specifically set up for psychical investigations. Encouraged by the ASPR to "try some experiments" with the promising medium, McComas insisted that Cartheuser conduct a sitting in his own private study. McComas felt assured that this professional and meticulously ordered space was certain to be free of any "tricks or gadgets," in apparent contrast to the cottages of Lily Dale.[111]

McComas took pains to organize the room so that he could obtain at least limited vision of the medium and employed a stenographer to sit in dim light on the other side of a black curtain to record everything that occurred.[112] Intently observing the medium in this much more ideal space for scientific study, McComas described the methods Cartheuser used to produce the various voices with the trumpet. McComas detailed how Cartheuser subtly directed the trumpet to produce the effect of "roaring from above." To generate the falsetto voice of one of Cartheuser's main controls – a little girl named Elsie – he would "vary the tension on the vocal chord" – a technique McComas believed to be "most exceptional."[113] The space thus afforded McComas the opportunity to redefine once-esoteric, immeasurable powers as observable bodily gestures and capacities on the part of the medium.

Challenges posed themselves to psychical researchers in their attempts to construct psychical laboratories, including unwilling spirits who refused to be restricted to a single room. Such seemed to be the case with many of the early experiments with Mina Crandon in Boston. At an early stage in the mediumship, phenomena began to occur well outside the seance room. During the evening of June 3, 1923, a levitating table that had been apparently possessed by the deceased mother of one of the sitters chased the seance participant out of the room, into a bedroom, and down the stairs. The spirit Walter performed only slightly less dramatic demonstrations a few months later, reportedly stopping the clocks all around the Lime Street residence and operating the Victrola record player on the ground floor in other instances. In fact, Walter's operation of the ground floor Victrola was significant; for a time, it served as his main demonstration of supernormal ability. Yet many visiting investigators, and even regular participants such as J. Malcolm Bird, expressed concern with the so-called "upstairs downstairs technique."[114]

When the Harvard group first began examining the Margery phenomenon in 1923, the problematic nature of such a technique became especially clear. The investigation by William McDougall and his colleagues commenced shortly after the phenomenon began to range throughout the house, a situation that their preparations reflected. Before the sitting, the men searched the entire dwelling: "all closets opened and scrutinized, all beds looked under, trunks, chests, bureaus, etc., opened, sofas inspected." McDougall searched and locked both the grandfather clock and the Victrola, and he took personal possession of the keys. They locked all the external entrances of the Lime Street residence and sealed the cellar door with wax.[115]

Despite the added security, the investigators' concerns were not resolved. As Bird himself admitted, "the phenomenon had to be pursued about the house by the investigators. It was idle to attempt any disguise of the fact that, on general principle and from a critical standpoint, this was highly objectionable." Even barring the fundamental lack of control of the medium under such circumstances, the house itself had "an architectural complexity" that contained several "curious closets, crannies, cubbyholes large and small, blind shaft ways, etc." The terrain required that one be exceptionally experienced with the house in order "to acquire confidence in one's ability to search the premises adequately, to seal one part of the house against the invasion from another, or the entire structure against invasion from without."[116] The impracticality of the situation, followed by accusations of fraud, finally led to the conclusion that the setting of the seance had to be significantly reorganized. After the Crandons and Bird conferred on November 15, 1923, experiments with Mina Crandon were conducted "entirely in one room, in Lime Street or elsewhere."[117] Psychical researchers such as Le Roi Crandon and Bird recognized the significance of physically situating their science to ensure that their claims to scientific authority and credibility were "embodied spatially" – to borrow the insights of historians Crosbie Smith and John Agar. Restricting the experiments to what could be recognized as a "privileged scientific site," they explicitly set apart their investigation of the paranormal from the rest of the domestic space.[118]

Holding experiments within a single room "set apart" was only one aspect of the equation, especially as investigators became increasingly vigilant about creating laboratory spaces to conduct their science of the paranormal.[119] As Carrington asserted, "we need psychic laboratories – to study such manifestations and discover if possible, their underlying laws and mechanisms."[120] However, this task was far from simple, especially considering the long lineage of the seance as a primarily domestic, and thus feminine, space.

Some of the first spiritualist seances of the 1850s took the form of "home circles," which subsequently became the foundation upon which spiritualism developed from the mid-nineteenth century onward. Consisting of three or more people, preferably of mixed sexes, these circles were traditionally composed of close family or associates. Sometimes solemn, at other times entertaining, they emphasized the place of the seance within the home, of which woman remained the chief guardians. As scholar Ann Braude has identified, nineteenth-century spiritualist practice inherited much of its form from the "cult of womanhood," which envisioned women as the prime moral guides within the domestic sphere. Echoing this belief in womanly sanctity, spiritualism, according to Braude, "secured the place of religion well within the woman's sphere by relocating religious practice from the church to the home." Not surprisingly, then, the "ritual of the seance perfectly suited the domestic environment."[121]

As psychical investigators took a firmer grasp on seance practice from the late nineteenth to twentieth century, these men took it upon themselves to rectify this situation by ridding the seance space of all stereotypical feminine markings. To do so, they emptied the space of anything resembling domesticity, such as artwork, familial memorabilia, and plush furnishings. Psychical investigators reorganized the seance into an austere, laboratory setting and thereby projected their own authority as icons of the masculine scientific self onto the space itself. Paralleling sociologist Henri Lefebvre's analysis, psychical investigators forged the space of the psychical laboratory to be "a tool of thought and of action," which performed as "a means of production" as well as "a means of control."[122]

Recognizing the importance of erecting a space that conformed to scientific standards, the Crandons took further steps to convert an upper floor lounge of their Lime Street residence into a psychical laboratory. Stripped of plush furnishings and memorabilia, the room was equipped with a red light controlled by a rheostat, a simple wooden seance cabinet, a mahogany chair for the medium, a seance table, and various pieces of technical equipment required for the experiments. Their efforts did not mean, however, that the conditions the Crandons created were free from criticism. In December 1924, shortly after the *Scientific American* committee had ended its investigations, Eric J. Dingwall arrived from London to continue his investigations of Mina Crandon that he had begun in 1923. Upon arrival, he intimately associated himself with the Crandons' experimental room, which he meticulously measured and mapped before beginning his experiments. Despite such efforts, Dingwall still found the space problematic. The Crandons had not removed enough unnecessary furnishings, according to Dingwall. Even more troubling, the room was located in the medium's home – circumstances that he argued "robbed [the experiments] of so much evidential value."[123]

Attempts to create spaces set apart from feminine domesticity continued among other investigators similarly convinced of the need to transform the seance room into a psychical laboratory. Some of the early photographs of T. Glen Hamilton's telekinetic investigations vividly illustrate just this sort of shift, as the place of his experiments was transformed from an obviously domestic seance room into a psychical laboratory. Initially, Hamilton held his psychical investigations within the parlour of his home, which appeared to include many of the common features of other domestic environments of its kind. Through its abundant furnishings, photos, portraits, and memorabilia, the living room of middle-class homes commonly presented values of maternalism and the family to a limited public. Marked with typically feminine crafts of embroidery and decor, the space asserted women's position as the moral and spiritual authority within the home.[124] The living room within which Hamilton held his first psychical examinations followed this same pattern. Three photographs depicting his early experiments centrally display the hearth just behind Hamilton and the medium – a symbol of warmth, nourishment, and familial mutuality – surrounded by ornate decorative lamps, family photos, dried flowers, and floral printed curtains. A piano notably stands in the background, a prominent emblem of not only the home but feminine gentility and cultivation as well. An unnamed and unacknowledged woman sits at the stool of the piano, emphasizing the feminization of both the musical instrument and the entire space (see Figure 2 for the first of these three photographs).[125]

The aesthetics of this seance appeared to mirror spiritualist practice since the 1850s, arguably resembling many other women-led home circles. Yet Hamilton's description of the photograph in Figure 2 provides a much different perspective of what the image was intended to convey. Labelled as "First Experiment," the photo depicts the medium Elizabeth Poole being tested for her telekinetic power. The caption hardly mentions her, however, focusing primarily on the psychic force and the investigators

FIGURE 2 "First Experiment," March 26, 1923. A view of one of the initial formal experiments conducted by Dr. T.G. Hamilton and one other co-experimenter with the medium Elizabeth Poole in the parlour of the Hamiltons' home in Winnipeg, Manitoba.

who grappled with it: "Here the two investigators held the table and prevented it from moving, although it is lifting and straining to do so under their combined grasps," reads the back label of the image. Such a description foreshadowed the masculine-centred, scientific methods that eventually encompassed the entire space in which the experiments took place.[126]

From the second half of 1923 onward, Hamilton moved his experiments to a second-storey room. Unlike the parlour, this upper floor room could be much more easily monitored and regulated. Hamilton alluded to this added level of control when describing the room as "set apart for the purposes of metaphysical research," and he alone equipped and managed it. Hamilton could control access to this room, as well as more easily survey and monitor mediums once the room was emptied of unnecessary furnishings that might obscure his view. As the record of the seances insisted, "the medium was under his personal scrutiny from start to finish of each and every experiment."[127] In accordance with this added level of scientific control and objectivity, the aesthetics of the new seance space shifted dramatically from the previous one. This room contained little to no decor and minimal furnishings. In the place of curtains, plain paper covered the windows to restrict the entry of light. The circle of friends and family disappeared as the medium sat surrounded by male observers (see Figure 3).

Such a transition reflected a determination to produce evidence of repeatable paranormal phenomena in a standardized and ordered space. Harry A.V. Green, in his foreword to T. Glen Hamilton's *Intention and Survival* (1942), wrote that psychical experiments needed to be repeatable "*ad infinitum* by any number of experiments" so that "the same result can always be obtained." This requirement necessitated an established set of replicable spatial conditions that would lead to the effective production of paranormal phenomena.[128] Thus, rather than conceiving of the experience of the seance as unpredictable and uncanny, predominantly male investigators aimed to refashion it as a controlled experiment conducted within a rigidly ordered space under their watchful and objective eyes.[129] This reorganization led to the creation of a performative space for psychical researchers, within which they could challenge critics who insisted that the paranormal could not be regulated. On the contrary, psychical experiments could be duplicated again and again, much like any other scientific experiment. Restructuring the space in order to transform it from seance to laboratory, psychical researchers reconfigured communing with the spirits as a scientific exercise. Doing so required surmounting numerous challenges, as neither the mediums nor the spirits always

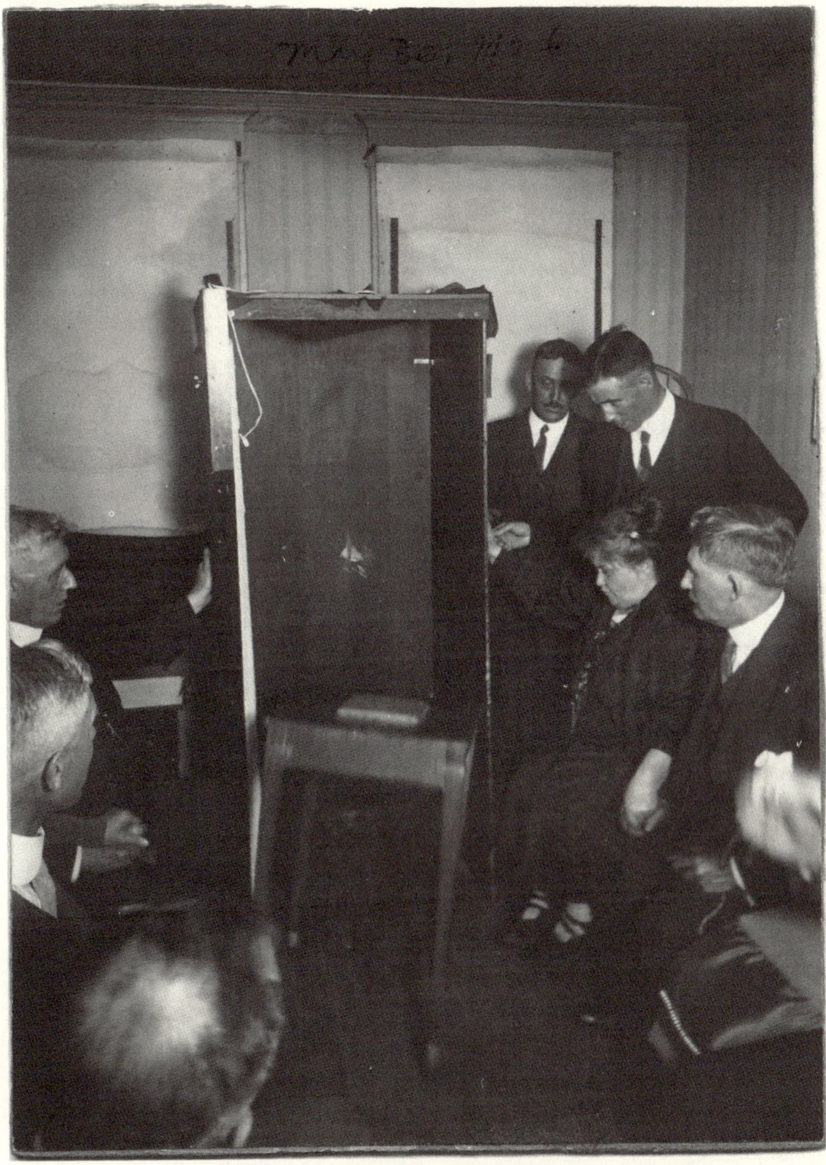

FIGURE 3 "Levitating Table," May 30, 1926. Photograph of a later telekinetic levitating table experiment with the medium Elizabeth Poole in the upper room laboratory of the Hamiltons in Winnipeg, Manitoba. T.G. Hamilton, Dr. Ross Mitchell, and Dr. J.R. Davidson surround the medium and hold her hands.

complied. Investigators such as McComas, Crandon, Dingwall, Bird, and Hamilton nevertheless persisted, cajoling mediums from their more entertaining seance rooms into regulated spaces, restricting spirits to one room, and restructuring the space itself so that it reflected the scientific standards these men so fervently championed. Within these psychical laboratories, they performed a scientific self that paralleled that of the natural sciences – objective, rational, and masculine.

***

Through their language, methods, gestures, and acts, men such as W.R. Wood, T. Glen Hamilton, Le Roi Crandon, Eric J. Dingwall, J. Malcolm Bird, and Henry Clay McComas performed the role of the impersonal experimenter. This enactment may often have been embodied, but it was intended to emphasize a researcher's tenacious use of the mind. Asserting their skills of precision, sense of objectivity, and disembodied rationality, these men reinforced an identity that aligned them with a much broader scientific community. Donna Haraway portrays the man of science as one who was "endowed with the remarkable power to establish the facts." This power translated into the very makeup of the person himself: "He bears witness: he is objectivity; he guarantees the clarity and purity of his objects."[130] Similar to Daston and Galison's description of the "scientific self," psychical researchers attempted to represent themselves as able to conduct empirical investigations of an extraordinary kind.[131] Collaborating with other like-minded investigators while building transnational networks of predominantly masculine investigators, they aimed to bolster their identity as legitimate scientists. Despite the contribution of several women to experimental procedure, research, and recording, the role of psychical scientist remained chiefly the domain of the masculine. Feminine bodies and minds may have been pivotal to carrying out investigations, but only the manly psychical researcher retained the intellectual and physical powers to design, conduct, and witness credible experiments. As investigators constructed their own sense of self, so too did they shape the environment of the seance. Psychical researchers transformed the seance from a mysterious, stereotypically feminine, and domestic space into an austere, standardized laboratory. In this fundamentally masculine realm, they performed their identity as the scientist to regulate the most prominent factors that could skew results and dismantle the carefully arranged experiments: the medium and the spirits.

2

Otherworldly Subjects: Mediums and Spirits

The essence of real scientific research is control of the influencing conditions.
– Waldemar Kaempffert, "The Scientist"[1]

Donna Haraway has argued that the ideal of the scientific man "inhabited the culture of no culture."[2] As the previous chapter details, interwar psychical researchers endeavoured to take on just this characteristic. Donning a "scientific self" typified by objective rationality, while remaking spiritualist seance rooms into laboratories, investigators aimed to produce new knowledge about the paranormal that was untainted by social bias or embodied subjectivity. Investigators argued that they could arrive at universal truths of the paranormal by enclosing mediums and their spirits within rigidly ordered rooms set apart from the world around them. Yet the interwar world refused to be shut out from their experiments.

Since the nineteenth century, the mediumistic subject had been conceived of as feminine, naive, impressionable, intuitive, and spiritually sensitive. Scholars Alex Owen, Ann Braude, Diana Basham, and others have observed that these qualities reflected Victorian feminine ideals, marking women as especially suited to engaging the supernatural.[3] For psychical researchers, the passive bodies of female mediums were exceptional instruments through which the spirits could act and be tested. Investigators prized psychics they perceived as easily malleable to empirical pursuits – compliant, uneducated women of a lower social rank.[4] Yet not all mediums conformed to this category. Women mediums did not always adhere to the stereotypes placed upon them. Men took advantage of opportunities to channel the otherworld as well, transgressing gendered dichotomies at work in the seance. The ghosts that emerged from these mediums only exasperated a problematic situation. Although the spirits provided a bridge between this world and the next, they demanded boisterous singing, cracked uncouth jokes, and donned racialized identities that disrupted the fragile seance conditions. As Molly McGarry and Marlene Tromp have illustrated, clashes like these characterized nineteenth-century spiritualism, as mediumship offered the possibility for men and women to evade the limits placed upon the gendered body.[5]

These struggles persisted into the twentieth century, continuing to provide "a barometer for much of what was occurring in the political and social world" beyond the seance room walls, as Tromp has asserted.[6] As this chapter argues, however, their form was fundamentally different from that of their Victorian incarnations, articulating the anxieties and tensions of interwar society. Efforts to contain subversive mediums and their unruly spirits paralleled wider fears surrounding the "modern girl," masculine degeneration, working-class rebellion, and racial assimilation. Rather than removing mediums from their surroundings, the psychical laboratory animated the cultural conflicts of interwar society. These conflicts were made tangible through the embodied performances of those mediums and spirits who refused to be passive bystanders to the experiments of the seance, engaging their investigators in unexpected, even troubling, ways. Their riotous behaviour ultimately challenged investigators' claims to empirical authority, undermining the larger project of transforming interwar psychical research into a detached and objective science.

Sensitive Creatures: Women Mediums and Queering Spirits

Conceived of as submissive, docile, and especially receptive to the spirits, female clairvoyants allegedly epitomized the compliant, sensitive specimen of psychical research. By filling this role, women hearkened to nineteenth-century spiritualism, which envisioned the medium as encapsulating the Victorian mores of feminine vulnerability, emotionalism, intuition, and obedience.[7] Changing conceptions of womanliness in the interwar period nevertheless had the tendency to seep into the seance room, revealing that psychical research was much more enmeshed within the twentieth-century world than even investigators cared to admit.

From the nineteenth century onward, women were viewed as more able to channel the spirits through their ability to utilize an extrasensory faculty. Known as the "sixth sense," or "cryptesthesia," as Charles Richet called it, this endowment was defined as "a perception of things by a mechanism unknown to us of which we are cognizant only by its effects."[8] It remained unique from the other senses in that it did not require the physical sensory apparatus and in fact could be impeded by external sensory stimuli. Researchers such as T. Glen Hamilton nevertheless argued that cryptesthesia "must be considered as a faculty in the medium just as definitely as we consider any other generally recognized sense."[9] The sensations it produced were experienced as "something quite real" but only by individuals who were first able to submissively lay aside their other physical perceptions, primarily through entering trance.[10]

Of all the various states of mediumship, trance accentuated the psychic's passive femininity, its resemblance to death reaffirming her role as a submissive body to be used by the spirits. Hamilton explained that the production of the most complex phenomena required that the medium be in trance, which he described as "a sleep much more profound than ordinary sleep, and showing many features of an abnormal type – loss of normal consciousness, anaesthesia of the skin, disturbances of the heart and respiratory rate."[11] When observing the medium Susan Marshall, or "Mercedes," as she was called in the seance, he remarked how her trance could at times resemble a "cataleptic state." Muscle and joints "became spastically fixed," with "the pulse slow and laboured, breathing reduced almost to imperceptibility."[12] Like catalepsy, trance rendered the medium completely incoherent, unable to control herself or the conditions placed upon her. If she did exhibit any activity during trance, it was proposed that the medium "function[ed] as an automation" rather than acting on any impulse of her own.[13]

Catalepsy has frequently been linked to or even mistaken for death.[14] Likewise, when mediums were deep in trance, their bodies were commonly described as a "corpse" by none other than the ghosts who spoke through them. A spirit speaking through the medium Ewan remarked of Mercedes, "Oh, she is dead. What are you going to do? You are going to get it in the neck. Do something quick. Don't sit there and stare. She wants something to drink." Hamilton asked, "Who does?" The spirit replied, "The corpse."[15] The spirit Walter in Boston similarly referred to the medium in deep trance as being "dead" and described Mina Crandon as "a beautiful corpse."[16] By asserting the lifelessness of the mediums, these spirits signalled their own ability to transgress death. They, not the mediums, were responsible for the vibrant psychical phenomena that erupted in the seance room. The medium's passive body was thus once more acclaimed but perhaps with purposes different from what psychical researchers had in mind.

Understanding female mediums as submissively offering themselves to be used by another had a long lineage that stretched back into the nineteenth century and earlier.[17] Alex Owen was one of the first of many scholars to recognize that the ascribed qualities of mediumship remained embedded within a complex gendered framework upon which the seance operated. The ideal medium was characterized as sensitive, impressionable, weak-minded, and perhaps most importantly, passive. Self-renunciation rested at the heart of what it meant to be a spiritual medium, which was also the necessary quality of the ideal Victorian woman.[18] Although seances underwent several transformations in the twentieth century, this fundamental characterization of mediumship remained relatively intact. Following the First World War, however, this very representation of womanhood was perceived as under siege.

In interwar society, the assumption of womanly submission to patriarchal authority was being drawn into question. Perhaps even more alarming than the "new woman" that emerged at the turn of the century, the "modern girl" most closely associated with the 1920s pointed to radically shifting mores around women's economic independence and sexuality. This modern incarnation of womanhood literally appeared across the globe, according to historian Alys Eve Weinbaum and others, to be internationally recognized through an assertive engagement with commodity culture and "explicit eroticism." Contemptuous of the Victorian roles of submissive and dutiful daughters, wives, and mothers, the modern girls of the interwar period challenged patriarchal authority to "define themselves in excess of conventional female roles."[19]

A number of scholars have questioned the supposed emancipation of women in the 1920s, asserting that this more modern, independent woman attained a symbolic rather than actual significance.[20] In fact, the term "modern girl" more often described delinquents, as opposed to liberated women. Yet, as scholar Tamara Myers argues, this alignment came about due to young women's perceived flouting of gendered norms, including feminine domesticity and sexual innocence.[21] Historian Mary Louise Roberts maintains that in the context of interwar France, the vision of the modern women indicated how change was conceived of in this period and thus cannot be ignored.[22]

The media quite possibly did exaggerate women's independence in the interwar era. That this public rhetoric surrounding women's newfound economic and sexual freedom became so prevalent internationally nevertheless reveals how shifting cultural mores and the anxieties they unleashed were understood and internalized.[23] Regardless of accuracy, this imagery may very well have had a powerful resonance among the particular network of psychical researchers under study, especially considering the urban centres in which it operated. Boston, for instance, where Mina Crandon lived and conducted the majority of her seances, was home to one of the largest urban populations of young female labourers for several years leading up to the period. Several twentieth-century reformers viewed this group of women, much like the modern girl, as far too independent, morally questionable, and inclined to transgress middle-class notions of respectable womanliness.[24]

Such anxieties were only compounded by the fact that in several ways Mina Crandon, or "Margery," seemed to embody this image of the interwar modern girl. With bobbed hair and a vibrant sense of humour, she was deemed "too attractive for her own best good," and her assertive sexuality troubled many who entered her seance room.[25] Her beauty, wit, and charm hardly went without comment in many of the reports of her mediumship; even her critics described Margery as "vivacious" and "a very attractive blonde, with a charming expression and excellent figure."[26] Margery's alleged ability to use her body to seduce men into conspiring along with her was often cited by her detractors. Her sensuality, "sufficient to disarm any mere man," easily led to infatuation, which was apparently enough to compel her male investigators to act as accomplices, or at the very least to ignore her acts of fraud.[27]

Joseph B. Rhine and his wife Louisa Rhine were two psychologists trained at the University of Chicago who expressed just such reservations about Margery. The Rhines have historically been credited with eventually redirecting the science of the paranormal. In 1927 they would join William McDougall, who had by then moved from Harvard University to become the head of the Psychology Department at newly founded Duke University. Appointed as faculty, Joseph B. Rhine would establish a new field of "parapsychology" with the support of Louisa Rhine, complete with a laboratory and academic personnel.[28] Their contributions were in many ways shaped by their first encounter with Margery only a year before their move to Duke. Invited by McDougall, the Rhines attended one of the Margery seances in July 1926.

Intrigued by paranormal phenomena, the Rhines expected a great deal from the seance. Yet both Joseph and Louisa complained about Margery's manipulation of those who were supposed to be her investigators. Margery's objectionable behaviour, including "kissing and embracing" one of her "more ardent admirers" during the sitting that the Rhines attended, made objectivity on the part of her examiners impossible. "Could this man be expected to detect trickery in her?" they asked rhetorically. Alluding to the "moral history" of the case, the Rhines claimed this flirtation on the part of Crandon was "by no means an isolated incident."[29] Beyond any concrete evidence of fraud, Margery's sexual licence inside and outside the seance room marked her as wielding far too much agency and power to be considered a reliably passive subject for psychical research.

That her main spirit control was a vigorous young man only intensified the subversive nature of Margery's mediumship. For those who affirmed Margery's ability, the performative manliness of Walter offered further legitimacy to the experiments through various demonstrations of the spirit's sexual distinction from his female medium. "Walter's treatment of the objects handed out for identification is always brilliantly consistent with his claims of personal identity," Le Roi Crandon declared in 1926. "Thus in dealing with a small barrette, he said that it was from a lady's hair, but not a comb – the failure to name it being decidedly a masculine touch."[30] Walter's flagrant heterosexuality, evident in his "prolonged whistle of appreciation" for any attractive woman who entered the seance, was further proof of his masculinity.[31]

Margery's assumption of a masculine character did not sit well with everyone. To some, it was one more indication of her sexual aberrance and tendency to undermine patriarchal authority. According to her critics, Walter was nothing more than a troubling impersonation of a man, a clever trick on the part of Margery to usurp the authority of her investigators – her husband included. As Henry Clay McComas amusingly observed of Margery's mediumship and the Walter personality in particular, "here was a unique way for a wife to manage her husband."[32] Margery's unwillingness to be passively controlled did not just threaten the credibility of her own psychic gifts. According to her most avid detractors, it placed the scientific ambitions of psychical research under threat. Viewing the Margery case as the "the greatest scandal in the whole history of mediumship," Joseph B. Rhine lamented its endorsement by the American Society for Psychical Research (ASPR), convinced that the society's unwavering support was counter to "the advancement of science" and would ultimately lead to the demise of psychical research as an area of study.[33]

Unmanly Channels: Male Psychics and Gender at the Margins

Margery joined many others who used mediumship to skew codes of normative gender and sexuality. Speaking of the nineteenth century, McGarry describes mediumship as "a kind of purifying transfiguration and release from the earthly gendered body," which in turn "held enormous appeal for women and men who inhabited gender and sexuality in transgressive ways."[34] However, men's assumption of supposedly feminine roles of mediumship did not go unscrutinized.[35] As Victorian-era mores gave way from the fin-de-siècle to the interwar period, shifting ideals of masculinity only intensified anxieties around men who communed with the spirits, anxieties that in the end proved detrimental to the scientific project of twentieth-century psychical researchers. William Cartheuser and Harry A.V. Green were two men who developed their psychic gifts during the interwar period. Despite taking on mediumistic roles, they countered the classification of effeminacy by either concealing their gifts from a wider public or presenting a much more aggressive form of mediumship. Nonetheless, their otherworldly powers threatened to skew the boundary between male and female, scientist and subject, while highlighting the turmoil surrounding gendered categories in this period.[36]

William Cartheuser established a reputation as a "remarkable medium" whose ability to compel spirits to speak through an aluminum trumpet and interact with the sitters confounded many.[37] He was inspected by a number of psychical investigators in the United States and also acted as the main medium for the Church of Divine Revelation in St. Catharines, Ontario, during the 1920s and 1930s. In each case, Cartheuser impressed both spiritualist and investigator with his mediumistic powers. He came from a humble background and was employed as a mechanic before beginning his career as a professional clairvoyant.

Investigators portrayed Cartheuser as they did several female mediums, describing him as "dull," "feeble minded," "simple," and "naive."[38] ASPR officials were first drawn to Cartheuser after reports he was channelling the famous Walter spirit in 1926, and they convinced him to hold a joint sitting with Margery in the fall of that year. Judging Cartheuser's Walter communications to be less than genuine, ASPR investigators assessed Cartheuser's impairment by "a hair lip" and "peculiarities of diction" as leading to an "inferiority complex" and an "apparent childishness of mind."[39] Even when praised for his gifts, Cartheuser's body and mind rarely escaped pathological classifications. Researchers such as McComas,

J. Malcolm Bird, and Hereward Carrington repeatedly referred to Cartheuser as "undersized" for a man, while claiming he demonstrated stereotypically feminine traits, including nervousness and apprehension.[40]

That one of Cartheuser's main controls for many years was the spirit of a young girl referred to as "Elsie" or "Alice" did not help matters. Known for her distinct voice – a "high childish soprano" – the spirit only further drew into question Cartheuser's uncertain masculinity.[41] Such categorization reflected deeply imbedded conceptions of the necessary femininity of any who would mediate between this world and the next. It also conformed to characterizations of the ideal scientific subject. As Evelyn Fox Keller has claimed, metaphors within the scientific establishment situated men as the rightful interrogators of a feminized nature, their main object consisting of "the illumination of a female interior, or the tearing of Nature's veil."[42] Cartheuser's male body represented a gendered paradox of not only mediumship but his place as a scientific subject as well.

Cartheuser's apparent adoption of a more wilful and vigorous mediumship by the early 1930s may have been in response to the gender politics he regularly faced. "Elsie" faded in importance from the late 1920s onward to make room for "Dr. Anderson" – a medical doctor with a "very powerful" voice, who was adept at diagnosing sick bodies and prescribing remedies.[43] Cartheuser's performance of mediumship changed during this time to one that reflected ideals of self-control and bodily mastery. Jenny O'Hara Pincock described one of Cartheuser's first visits to St. Catharines in almost heroic terms. Although she was "concerned that we might be taxing his strength," Cartheuser informed them "he wished to lose no time." Far from the timid, nervous person that officials of the ASPR described, Cartheuser was an emblem of endurance and stamina, able to conduct lengthy, elaborate seances without weariness.[44] Intense physicality, in turn, characterized his sittings, as objects whizzed in every direction, while spirits aggressively grappled with the sitters: "The terrific speed of the trumpets," O'Hara Pincock recalled, "swept the séance room with a sound as of a mighty rushing wind."[45] At other times, sitters reported that they were struck hard on the knee, "banged ... in the stomach," and "hit in the breast."[46] Dr. Anderson expressed an ethic of self-mastery throughout many of his mediated lectures, emphasizing that a "medium must be the main control over any entities, over spirit guides and teachers."[47] A year later, in 1932, the spirit doctor bellowed, "Remember, you are the Master of the Body."[48]

Concepts of manhood shifted dramatically between the 1880s and the 1920s. In response to radical social changes such as corporate mergers, immigration, working-class activism, and women's movements, many attempted to reform notions of the ideal man. Victorian manliness, characterized by self-control, intellect, and chivalry came under fire during these years, to be potentially superseded by a new form of virile, aggressive, and physical "masculinity" – a term that was redefined and employed to describe this new man of the modern era. Particularly in the United States, this ideal was structured in response to race relations and the perceived threat of black manliness. In the place of the "overcivilized" and sickly office clerk of the late nineteenth century, this masculine creature, characterized by strength of body, would rejuvenate the (white male) human race.[49]

Yet despite this push to solidify such gendered ideals, concepts of manhood remained unsettled at the end of the First World War.[50] In fact, the war only deepened a feeling of social and cultural upheaval, especially in regard to gender. Fear of the independent and promiscuous "modern girl" rose following the First World War, no doubt spurred by women's recently won enfranchisement, as well as their steadily growing entry into the workforce. Shifting marital and sexual mores only added to an already complicated situation. In North America a number of social commentators argued for an increased focus on companionate marriage and more egalitarian relationships.[51] Meanwhile, in Britain, men were encouraged to take a more involved and sympathetic parental role to enrich family life.[52] At the same time, returning soldiers, frequently emotionally and physically broken, threatened ideals of the respectable Victorian gentleman, as well as the muscular modern man.[53] Historian Joanna Bourke has argued that the war did not necessarily lead to a radical cultural shift in how masculinity was understood. It did hold many implications, however, for how the male body was perceived in the interwar period. No longer "the sum of its various parts," prevalent images of the deformed and disabled man returning from war emphasized the vulnerability of masculinity.[54]

Resisting classifications of pathology and effeminacy, Cartheuser attempted to present a mediumship that encapsulated more modern ideals of rigorous masculinity. Despite his efforts, the ambiguity that shrouded manliness in the interwar period, to which his body and mind seemed to draw attention, proved difficult for Cartheuser to escape. Even those who fully supported Cartheuser and believed him genuine did not find such displays enough to counter his allegedly frail body and childish, effeminate mind. Cartheuser's growing dependence on O'Hara Pincock led her sister, Minnie Maines, to report, "William is like a lost sheep without you."[55] When he attempted to establish independence from O'Hara Pincock in the mid-1930s, Cartheuser still found himself characterized as impressionable and dependent – if not upon the spirits, then upon the sitters he wished to please. According to O'Hara Pincock, Cartheuser increasingly "coloured" his messages, "just to suit the minds of people he wished to gain favour with."[56] Reflecting on her experiences with the medium in the 1940s, she claimed that he too easily bent to the wishes of others rather than being steadfast in his spiritualist convictions. As opposed to the self-mastery he attempted to demonstrate, Cartheuser revealed himself to be weak-willed, a temptation "most mediums succumb to," O'Hara Pincock lamented.[57]

Harry A.V. Green similarly negotiated the uneasy gendered position in which he found himself as a result of his psychic inclinations, concealing and even reforming the manner in which he engaged with the spirits. Green became involved in the Winnipeg experiments of T. Glen and Lillian Hamilton in February 1928. Educated in law at Edinburgh University, he came to Canada in 1913 to establish a practice in association with the Canadian Pacific Railway.[58] T. Glen Hamilton described Green as "a man of education, culture and the highest integrity." "Possessed of a highly critical mind and a more than ordinarily cautious outlook," Green did not appear to fill the necessary mental or emotional requirements of submissive mediumship.[59] Initially, he joined the Winnipeg circle as a "sitter and observer and with no thought in mind of developing mediumship either physical or mental."[60] Despite such motivations, he soon began acting as an unwitting channel for the spirits with impressive demonstrations of trance that rivalled the female clairvoyants in the seance.

Green nevertheless seemed uncomfortable with this role. Hamilton noted that although Green's mediumship proved invaluable, he "found it more difficult than most to allow his psychic gifts to be developed, and when they were developed, to regard their products as of any particular value."[61] The spirits, in turn, complained, "I can use this boy, but not well."[62] Refusing to allow his name to be used in either Hamilton's publications or experimental notes, Green first went under the designation of "X," which the spirits changed to "Ewan" in September 1929.[63]

Green's, or Ewan's, performance also differed from that of his female counterparts in the seance, as he asserted a much more aggressive form of mediumship. Such tactics may not have been as unusual as one might think. Historian Joy Dixon has identified that occultism was increasingly linked to male virility in Green's British birthplace during the 1910s and 1920s. Unlike "mysticism," which was characterized as passive and thus feminine, "occultism was based on the cultivation of a magical willpower" and thus "coded as masculine."[64] Green did not take on feminine ghosts like Cartheuser but let himself be controlled by the spirit pirate John King – a popular and rowdy incarnation of the seventeenth-century buccaneer Henry Owen Morgan – and occasionally by other male-identified ghosts such as Walter.[65] Barking commands and even intimidating the other mediums while in trance, Ewan seemed anything but demure. "Be quiet!" he snapped at the medium Mary Marshall, or "Dawn," as she was called in the seance. When she refused, the seance records reported that Ewan "[stood] up over Dawn and silence[d] her."[66] Noted for "frequent spasmodic movements of the limbs," in addition to "occasional muscular contractions" and "exceedingly rapid breathing," the entranced Ewan was vigorous and active – "an energy-producing centre not only at the end but through the whole series of experiments."[67]

Hamilton and his colleagues, however, did not interpret such dynamic performances as a display of vigorous masculinity. In January 1930 the records indicated that Ewan again commenced an almost athletic performance when entranced, "clapping hands and stomping about" while standing atop the seance table. The record keeper for that night, W.E. Hobbs, made a note of it, classifying Ewan's display alongside the various behaviours of the entranced Mary Marshall as "a period of nervous excitement."[68] In addition, Ewan's lively performances not infrequently gave way to a much more passive and vulnerable trance state, his "respirations" becoming "extremely subdued," only to be "resumed with great effort and deep-drawn inspirations as if the respiratory muscle itself were in a state approaching spasm, associated with whimpering, panting and occasionally violent sobbing."[69] The records for November 29, 1931, indicated that Ewan uttered "pants, gasps and sobs," followed by collapsing onto Lillian Hamilton's lap, where he lay demurely for some time in trance (see Figure 4).[70]

As a man communing between this world and the next, Green not only disrupted long-held ideals of feminine clairvoyance but also subverted the deeply ingrained gender dichotomies of the scientific enterprise, which psychical scientists increasingly attempted to assert within the seance room.[71] Skewing the boundaries between the masculine knower and the elusively known feminine, Green's mediumistic power moved between the polarities of male and female. He still maintained his position as an authority of psychical investigations and even wrote the foreword to T. Glen Hamilton's posthumously published account of his experiments, *Intention and Survival*.[72] But this status was only possible while

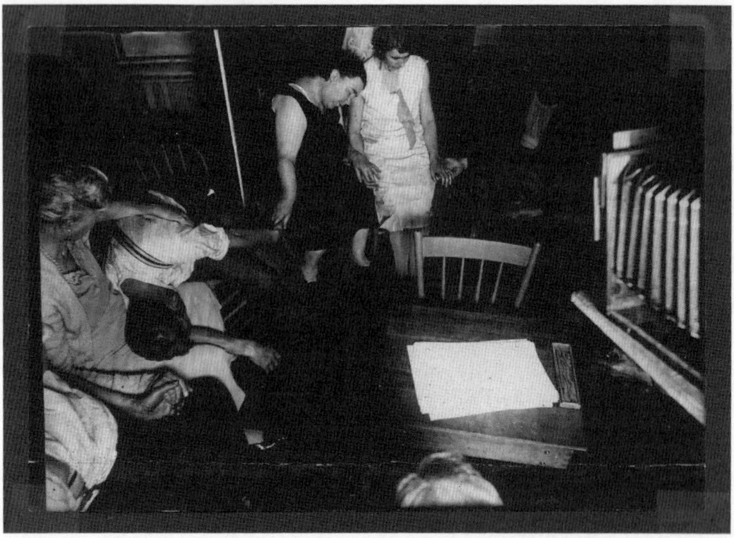

FIGURE 4 "The Second Lucy Materialization: Angle View of Experiments of November 29, 1931," a photograph within an annotated album belonging to the Hamiltons, featuring an experiment conducted in their psychical laboratory in Winnipeg, Manitoba. Mary and Susan Marshall hold hands with the alleged "Lucy" materialization behind them. The two women sit beside the medium Ewan, who has collapsed into Lillian Hamilton's lap.

his identity remained concealed. Green maintained his anonymity in all public records, insisting upon the use of his pseudonym of Ewan for years after the Winnipeg experiments had come to an end.[73] It seems both Green and his fellow investigators recognized the transgressive nature of his acts within the seance, which his display of a more aggressive masculinity could not completely conceal.

"Manliness is a cultural site that is always under construction," historian John Kasson has argued.[74] Both Cartheuser and Green endeavoured to shape their masculinity in the seance room by reforming the terms of their own mediumship. As they struggled to disrupt the category of feminine mediumship, their assertion of bodily self-control and aggressiveness reflected a vigorous manhood – one well suited, as Kasson has suggested, "to face the challenges of modern life."[75] Their efforts were not always fully convincing, as they exposed the doubts surrounding conceptions of interwar masculinity. These doubts, in turn, threatened the psychical researchers position of gendered, scientific authority, which the spirits only amplified.

Working-Class Hauntings: Mediums and Spectres of Another Rank

Transgressing gendered and sexual mores was not the sole means by which mediums and the spirits proved problematic to the ambitions of interwar psychical researchers. In addition to being associated with the feminine, it was not uncommon for mediums to be characterized as working-class, even though this was not always the case. This representation of mediums again anchored an idea of passive mediumship stemming from the nineteenth century, which safely placed lower-class women under the rightful control of middle- to upper-class psychical researchers.[76] Yet the labour struggles of the interwar period did not leave the seance room unscathed. In fact, upon closer inspection, the class of mediums and the nature of the phenomena they evoked in the interwar period signalled a much more troubling reality outside the seance room walls.

Although strikes and labour movements were infrequently mentioned within the seances under discussion, labour radicalism formed the context in which these seances occurred, shaping broader conceptions of both working- and middle-class consciousness. In Winnipeg the 1919 General Strike wreaked havoc upon the city, fuelling fears across the country of the potential breakdown of established class hierarchies. At least one participant in the Hamilton seances, Isaac Pitblado, was instrumental in the Citizen's Committee – a very discreet group of the city's elite who were credited with ending the strike and criminalizing the workers who dared to protest.[77] Numerous other Canadian industrial centres experienced class tensions and labour unrest, leading some historians to label the period 1917-25 the years of the "Workers' Revolt."[78] Similar pangs were felt south of the border in New York and Boston, as well as across the Atlantic in several British cities that experienced disruptive class conflicts and mass strikes. Beginning during the First World War, labour radicalism spiked in 1919 and continued throughout the interwar period. From clothing workers in New York to policemen in Boston and London, "No one seemed immune from the spirit of labour militancy."[79]

This labour activism was by no means uniform, especially across the Atlantic, as each movement responded to specific cultural understandings of class that emerged from different national contexts. Conceptions of who was included or excluded from such categories fundamentally shaped labour relations. In the United States, white workers actively attempted to limit, if not eradicate, the participation of black workers in unions to the point of undermining strikes and political action.[80] In

England the perceived need to exclude new immigrants, Jews, and a host of racial minorities so as to consolidate understandings of a unified working class, defined by whiteness and British birth, exploded into violent race riots in 1919 and beyond.[81]

Exclusions operated on the level of gender as well. As historians Joy Parr, Elizabeth Faue, Joanna Bourke, and others have asserted, radical labour activism in Canada, the United States, and Britain remained closely associated with the male worker.[82] Women workers at this time were also fighting for fair, standardized wages and better working conditions. However, the paternalistic manner in which government, employers, and male-dominated labour unions tackled such concerns revealed the persisting cultural restrictions women faced. Due to women's characterization as secondary labourers who entered the workplace only briefly before fulfilling their assumed destiny as wife and mother, patriarchal norms of pliable and dependent femininity continued to overshadow their attempts to identify as self-sufficient and capable workers deserving of fair wages and safe employment conditions.[83]

Compliant working-class women mediums arguably performed an affirming function for middle-class sitters. Their bodies laid bare for unrelenting scrutiny and speculation, these psychics held little power over psychical experiments in which middle- to upper-class psychical researchers took on the role of scientist to uncover the hidden secrets of a feminine nature.[84] The predominantly masculine image of radicals associated with the general strikes and labour activism of this period stood in definite contrast to the naive, working-class, female medium. Carefully controlled by her middle-class investigators, she embodied a likeness of the working classes and of women that securely maintained class and gendered hierarchies.

Psychical investigators generally viewed ignorance on the part of their women mediums as an important asset for scientific investigations, and they stressed this quality again and again. This requirement meant that the medium had to be either very young or of the lower or working classes – a woman who, as an effect of having minimal access to education, remained characterized as a "naive" participant with an "unsophisticated mental outlook."[85] Hamilton depicted the mediums with whom he experimented as women from humble origins and with limited access to education. Elizabeth Poole, Mary Marshall, and Susan Marshall were all working-class women, repeatedly described by Hamilton and others as having "few educational advantages" and as generally "unlettered."[86] Even Poole, who was a nurse and presumably had some degree of training, was portrayed by Hamilton as possessing only a "rudimentary education" and limited intellectual ability.[87]

Mary Marshall first became involved in the Winnipeg experiments due to an unusual set of circumstances. She had been known as a "crystal gazer," a gift that she used as a meagre source of income. In December 1924 she was employed to give a reading at the home of Eva Broad, sister of the law partner of Hamilton's brother, W. Oliver Hamilton. On this evening, Marshall correctly predicted the sudden death of W. Oliver Hamilton, which happened the following day.[88] When the Hamiltons were made aware of the premonition, they became intrigued and eventually convinced Marshall to take part in their experiments three and a half years later.[89] She related the circumstances of her upbringing to Lillian Hamilton in 1945, several years after T. Glen Hamilton's death. Originally from Govan, a working-class district of the city of Glasgow, Scotland, she described her family as "very poor." Her mother was a domestic servant and her father a farmhand.[90] As Marshall recounted, "the class of people where we lived were very poor, my father never earned more than 21 shillings a week."[91] Minimally educated, she entered the workforce as a domestic servant at the young age of eleven.[92]

On first appearance, Margery did not conform to the stereotype of the uneducated medium from the lower ranks of society. As the wife of a prominent Boston surgeon, she was comfortably positioned as a member of the upper middle class. Her supporters bolstered this identity. Referring to her as "Mrs. Crandon," they emphasized her respectable social standing, as well as marital status, which safely placed her beneath the economic and sexual control of her husband.[93] Since the late nineteenth century, such identity markers had remained important for mediums, who were often viewed as sensual, promiscuous, and amoral.[94]

Yet Margery's personal history revealed a social standing that was in fact much more ambiguous. Born on a farm in Picton, Ontario, in 1888, she was one of many migrant Anglo-Canadian workers in the early twentieth century who moved to urban centres in the United States to obtain employment. In 1904 she immigrated to Boston to take up clerical work, which she continued to do until she met and married grocer Earl P. Rand in 1910. After obtaining a divorce from him in 1917, she volunteered to drive an ambulance during the First World War for the very same naval hospital in Chelsea, Massachusetts, at which Le Roi Crandon acted as a surgeon. Here, they became romantically involved and married near the end of the war – a relationship that elevated her into the upper echelons of Boston society.[95]

Throughout the numerous publications that publicized her mediumship, Margery's more humble origins were rarely referred to. Described as charming, intelligent, and "very cultured," she in many ways stood in contrast to the stereotype of the uneducated, unprivileged medium.[96] Yet her class background did not disappear altogether as she became the channel for the spirit of her dead brother, Walter. When alive, Walter had been a common labourer, travelling from eastern to western Canada to work as a farmhand at age eighteen and later as a railway worker. While employed in this trade, he died in a locomotive accident at the age of twenty-eight. Although reportedly able to "tilt and levitate" tables while alive, such phenomena would be little match for the psychic feats he would accomplish once dead.[97]

Demonstrating a propensity for the production of psychical phenomena, the spirit of Walter levitated tables and various other objects, communicated complex messages, saw clearly in the pitch dark, and produced materializations that poured forth from his living sister's body. At the same time, Walter revealed an uncouth manner, a crass vocabulary, and a fondness for jazz music. Teasing and joking, often at the expense of various sitters, Walter whistled popular tunes, his favourite being "Souvenir," and he reportedly "kept time to fox-trot music which he emphatically preferred to more sedate, classical selections."[98] As he was known for his working-class, populist behaviour, even imitators resorted to crude humour and language to represent his ghostly presence.[99] Thus Walter, a personality that "dominat[ed] Margery's séance room," performed not only his own class identity but, perhaps, that of his medium as well.[100]

If Margery could not express her own class identity and frustrations, Walter had the freedom to do so. Arguably, Walter acted as a medium for another class of people whose access to the seance room remained limited and tightly constrained. His outrageous behaviour and demands, coupled with denigrating remarks, critiqued the social attitude of those in the circle. During a Boston seance in 1926, Walter told the sitters, "You're all snobs." Continuing, he insisted, "the street cleaner, the sewer digger or the kitchen worker may be more spiritual than the white-collared person who is so sure that he is God's representative."[101]

Walter was undoubtedly a raucous subject for psychical research, a trait that followed him far beyond the Boston seances. T. Glen Hamilton focused much of his earlier Winnipeg investigations on the discarnate personalities of famed historical figures such as William T. Stead, Robert Louis Stevenson, and David Livingston.[102] Much would change, however, when Walter began to present himself in the Hamilton seances in the form of a spirit control initially unrecognized as him and known only as the "Fair Young Man," or "FYM." Almost immediately, the environment became explicitly irreverent, with profanities, off-coloured jokes, and popular tunes. When this spirit first began to assert his presence in the Winnipeg seances, Hamilton failed to take him seriously, ignoring the Fair Young Man's requests and refusing to grant him any degree of authority.[103] The reason may have been the character and behaviour of the spirit, which did not parallel the expectations of the white, middle-class seance participants. When the Fair Young Man first emerged, he appealed to the sitters to refrain from solemn hymns. He asked instead that the group "sing something lively," recommending "O Dem Golden Slippers" and "Clementine."[104] Rather than an enlightened sermon, the first words the Fair Young Man spoke for all the sitters to hear were "I'll ring that damn box yet!"[105]

As opposed to sombrely complying with the gravity of Hamilton's scientific enterprises, the Fair Young Man's vibrant humour often caused the medium to laugh uncontrollably while relaying the spirit's jokes, insults, and crude language. During a seance on March 15, 1928, the medium Mary Marshall "[shook] with laughter," reporting that the Fair Young Man had mocked the sitters as being "so Scotch that the heather is growing out of our ears ... MM again in a spasm of laughter, says 'Oh!' he says, 'Damn it!' ... She remarks that FYM must be very quick tempered and ready with his tongue."[106] At times, the precocious spirit's joking came at the direct expense of the prestigious scientific personnel. When Hamilton announced the coming of noted archaeologist and prominent psychical researcher Bligh Bond, the Fair Young Man proceeded to make fun of the man's bald spot, threatening that he would "cop him on the head."[107] During Bond's visit a few days later, the mischievous spirit continued teasing and mocking. Proudly claiming to be "as bold as the devil himself," the spirit shouted, "Hello! Where is Baldy? Your B's. Bonnie, Baldie, Bligh Bond. He is not a bad fellow but is slow. I tickled the old boy's pate!"[108]

At this early stage of his involvement with the seances, the Fair Young Man often entirely overtook the mind and body of the medium to reveal otherworldly realities. Yet, rather than speaking of culturally heightened or scientific matters, the mysterious FYM used such moments to emphasize his background as a working-class railway worker in a comical and rowdy fashion. During one 1928 seance, Marshall entered a deep trance and began making "a noise like a train shunting." Once she emerged from the trance, she relayed to Hamilton and the other experimenters that she had been on a train with the Fair Young Man, who "was singing something about 'When I get there I'll see my girl!'" When they had arrived at the envisioned station, the medium witnessed him "wave to the people beside the gate" and "[shout] to two girls."[109]

Such antics directly contradicted Hamilton's insistence that these pursuits were a serious matter of strictly scientific interest, which needed to be limited to those who were professionally equipped to conduct the experiments.[110] The Fair Young Man's crude humour and working-class background branded him as an individual who was beneath the notice of professional, middle-class psychical investigators. But when the precocious ghost revealed his identity as none other than Walter, the dead yet vigorous brother of the famous Margery, Hamilton's attitude toward the spirit shifted radically.[111]

Hamilton had first been exposed to Walter during his 1926 visit to Margery's seances in Boston and considered his encounter an "important milestone in my study of metaphysics."[112] His experience with the Margery case deeply influenced the direction of his own investigations, especially after the Fair Young Man introduced himself to the Winnipeg experiments in 1928. Although hesitant to proclaim this spirit as the Walter of the Crandon experiments, Hamilton nevertheless insisted "that our Winnipeg Walter resembles in many ways, in character, vocabulary, in mannerisms, methods of work and types of phenomenon produced, the Boston Walter as I know him."[113] Such an observation did not seem completely unfounded, as both characters revealed a crass humour alongside marvellous supernormal power. If proven true, it would help to substantiate the existence of a ghostly personality independent of one particular medium that could not be the product of deceptive conjurers and associates.[114]

To align the unruly spirit of Walter with his empirical ambitions, Hamilton no longer conceived of the ghost's crude antics as frivolous acts in which he was only "casually interested" but regarded them as scientifically evidential occurrences. Rather than just a product of the subconscious of one medium, Walter's riotous behaviour was evidence of a transnational cross-correspondence between seances miles apart.[115] A few years later, in 1933, Hamilton did admit that the Walter entity was "a joking, teasing, breezy personality using the language and expressions common to young Canadian farmer lads and Canadian workmen." Hamilton quickly followed such a description with the observation that this spirit also remained "keenly interested in and apparently extremely anxious to uphold the scientific aspect of his phenomenon at every turn."[116] Such efforts obscured the ghost's origins but failed to dissolve them altogether.

Walter's demonstrations of an all-too-familiar affinity with the working-class mediums from which he emerged proved difficult to conceal. To make matters worse, Walter and other working-class apparitions were not the only subversive spirits with which psychical scientists needed to contend.

Playing Indian: Whiteness and the Dematerialization of Race

"Rhythmical stamping is heard as of an Indian dance," the Hamiltons recorded of a seance in November 1932. "Queen reports that it is Dawn. She is very heavy on her feet. She stops the dance and talks jibberish."[117] Such performances were hardly rare in seances during the interwar period, reflecting a long association between racialized ghosts and spirit communication. From the nineteenth century onward, "Indian" spirit guides frequented seance rooms, leading investigators such as Herbert J. Spindon of Harvard University to claim that spiritualism and Indigenous religious practices closely resembled each other. Speaking before the Boston Society for Psychical Research in 1928, Spindon asserted "that it is probably significant that spiritualism as we now understand it began in upper New York State near the reservation of the Iroquois Indians."[118] Communing with ghostly Indigenous incarnations was a means of "going native," to use the words of cultural theorist Shari M. Huhndorf. Seance sitters coped with the "tumultuous modern world" through their interactions with these racialized spirits, "idealizing and emulating the primitive, modernity's other."[119] Many who communed with the spirits insisted that Indigenous spirit guides granted white sitters access to spiritual realms that were inaccessible to allegedly more advanced and civilized seance participants. Given names such as "Bright Moon," "Trickling Water," "Eagle Feathers," "White Bear," "Bluehide," and "Blackhawk," Indigenous spirit guides typically spoke in broken English, their language laced with allusions to the natural environment.[120] A wild, natural figure of ghostly realms, the Indigenous spirit could aid overcivilized men and women in pursuing higher, esoteric truths within the seance.[121]

An attempt to capitalize on the place of Indigenous apparitions within seances ensued in the interwar period. The New York spiritualist retreat of Lily Dale presented itself as a natural and spiritual reprieve from the assault of civilized life. Constructing an "Indian Trading Post" on its grounds, Lily Dale administrators invited Indigenous performers such as

"Princess Nieoma" to provide daily entertainment for those who visited throughout the 1920s and 1930s.[122] Individual mediums at this time also invoked the spectre of the more natural and primitive Indigenous person to connect to the world of spirits. William Cartheuser was noted for his ability to compel these entities into action within the seance, the most prominent being "White Bear." This racialized spirit reiterated many of Dr. Anderson's mediated instructions and worked diligently to create the appropriate conditions for successful spirit communication.[123] White Bear's specific tribe remained vague, at times conveying a stereotypical representation of a "Plains Indian" and at other times insinuating an Inuit, or "Eskimo," identity – a representation of Indigenous culture that became exceptionally popular between the world wars. Blending myriad Indigenous cultures into one, the ghostly White Bear proved easily consumable.[124] Alongside other Indigenous spirit guides who were also present within the seance, White Bear primarily fulfilled the role of helper to both the white sitters and spirits. "Me see a nice squaw around you," White Bear commented to a sitter, "help you much."[125] Similarly, during a seance on November 26, 1928, a spirit guide called "Bright Moon" was asked, "Why do you have so many guides from your race?" Bright Moon answered, "Me know how to help you and draw earth vibrations."[126]

These spirits proved more satisfying than anything white sitters believed they would encounter in the real world. Although officially shown to be no longer "disappearing" by 1920, the conviction that First Nation populations were declining continued to be a prevalent idea throughout the interwar period.[127] The precise explanation of their alleged extinction nevertheless shifted to insinuate the disappearance of genuine Indigenous culture rather than Indigenous peoples themselves.[128] Anthropologists such as Marius Barbeau advocated the need to reorient efforts toward "salvage ethnography." This methodology was aimed at deciphering and preserving "authentic" Indigenous traditions and objects from "intrusive materials" that had supposedly supplanted traditional ways of life to the degree that, in his view, "most Native cultures in Canada were already dead."[129] Tactics such as these reinforced the belief in white dominance and racial superiority along evolutionary lines, thereby justifying segregation and exclusion while generating nostalgia and fascination with the more primitive and increasingly invisible other.

Indigenous peoples could thus still be transfixed into reassuring shades of ever-dissapearing racial identities, their ability to act contrary to white wishes being muted by ambitious government programs of cultural annihilation.[130] Within the Canadian context, revisions to the Indian Act in 1914 stipulated limitations upon the movement of First Nation people outside of government reserves, while instituting compulsory school attendance at government-funded boarding schools.[131] These revisions continued to be enforced throughout the 1920s and 1930s in an effort to legislate forcible assimilation. In 1920 the Indian Department's deputy superintendent, George Duncan Campbell Scott, stated clearly that the aim of such tactics was the elimination of Indigenous culture and identity. "Our object," Scott asserted, "is to continue until there is not a single Indian in Canada that has not been absorbed into the body politic, and there is no Indian question, and no Indian department."[132]

South of the border, violent pacification shifted to forcible assimilation from the 1880s to the 1930s, with the chief aim of destroying tribal community identity. As scholar Alan Trachtenberg has identified, US government officials claimed that such tactics were necessary "to prepare the Indians for U.S. citizenship, but also primarily to loosen for white ownership the lands natives held collectively." "Tribal form," it was believed, "prevented individual Indians from adopting the outlook (individualism) and skills (private ownership of land and goods) required for the proper civil identity of an American."[133] Philip Deloria argues that these governmental actions "eradicated" Indigenous peoples' supposedly innate propensity for violence, making them "invisible" even if they would not actually die off. On the one hand, these actions were intended to ensure that First Nation peoples would "melt into American society or sit quietly in the marginal distance, no longer disturbing anyone." On the other hand, they "rendered Indians safe, thereby opening up their lives and lands to visitations of various breeds of primitivists."[134] Moreover, twentieth-century white Europeans and North Americans embraced the idea that by "playing Indian" they could access the therapeutic primitivism that Indigenous peoples represented. Enacting their yearnings for a world they deemed more natural, they simultaneously asserted their own difference from the racial other.[135]

Seance participants likewise entered into "Indian" play through the medium's body, envisioning entities that fulfilled all their desires and racial presuppositions. Yet, as useful as these spirits could be for reaffirming white identity, they were not always welcomed, especially by psychical scientists who expressed a definite preference for spirits who would not be so unruly. Henry Clay McComas lamented in 1935, "One meets altogether too many Indians in séances." For him, the voice of such spirits alone proved far too disruptive to seance conditions. Whereas he applauded the oral gifts of several of Cartheuser's more identifiably white spirits, McComas complained of the spirit "Black Bear," who abruptly "dropped in."[136]

McComas was hardly alone in his grievance. In fact, even other spirits, or at least those who identified as white, expressed similar resentment. Walter, the "Fair Young Man," insisted that exotic spirits could be detrimental to seance conditions if not carefully controlled – a truly ironic claim considering how riotous Walter himself could be. Arguing that the "Indians" were "enemies" of reliable psychical investigations, Walter, speaking through Mary Marshall, complained that these typically unruly spirits "come in large numbers and they like to make fun. There is a great number of Indians who don't know anything at all ... They interfere a great deal with what we are trying to do."[137] As a result, these spirits needed to be carefully regulated. On October 17, 1929, through the medium Cartheuser, the spirit Dr. Anderson warned, "If there are any Indians come through, hold them to their reins. Do not let them carry you away. Hold them down with your thoughts ... Your thoughts can hold and control him."[138] Despite the aid provided by both White Bear and many other Indigenous spirit guides, they apparently retained their potential to upset seance conditions and even harm the psychic. Dr. Anderson argued that one should avoid deep trance states in case an Indigenous spirit came through, as their very presence could prove "dangerous" to the health of the medium.[139]

The Hamiltons had a long history with Indigenous spirit guides. One of T. Glen Hamilton's initial exposures to paranormal phenomena happened in association with Rev. D.N. McLaughlin of Winnipeg. Both intended to investigate the theory of telepathy but ended up experimenting with something altogether different. McLaughlin and Hamilton participated in a series of seances during the autumn of 1918 with a certain "Mrs. Ross" and a young medium named Lucy. Lucy exhibited a number of different spirit personalities, the "most astonishing" being a spirit who "claimed to be an Indian named 'Bluehide.'" The spirit spoke "in a sort of 'broken' English, a form of speech utterly foreign to the little child and of which she knew nothing." As an example of such foreign speech, Lillian Hamilton recorded how Bluehide's interpretation of the name "Lillie" involved a long description of "the lily pad that floats on the water," which "conveyed the central idea of the word in a manner foreign to the child and by the observers totally unexpected."[140]

Many years later, the participants of the Hamilton seances were contacted by a spirit guide that would significantly influence the future direction of the sittings. He first appeared on the night of April 11, 1928: "MM sees a magnificent Indian standing in the centre of the room. His arms folded; he has a lovely countenance. MM entranced." After he had introduced himself as "Black Hawk," his first words to the sitters made his role in the seance clear: "Me help you," he claimed. "Me not forget to help."[141]

Unlike most of the Indigenous spirit guides at these seances, Black Hawk was an identifiable historical figure. Responsible for mounting a fierce conflict against the United States in 1832, or "Black Hawk's War," as it was known, he became even more famous after he was forced to participate in his own "prisoner of war tour," ordered by president Andrew Jackson in the same year.[142] Dying in 1838, he began to appear in spiritualist seances from the 1860s onward, along with myriad other well-known Indigenous figures of resistance.[143]

The spirit who made his presence known in the Hamilton seances still seemed ready to contest white authority but in a way that reaffirmed typical representations of Indigenous violence that had been established by white colonists and settlers since the nineteenth century.[144] Always on the verge of aggressive upheaval, this spirit and his mysterious "fierce-looking" accomplices posed an ever-present danger to the delicate conditions necessary for successful psychical phenomena. During a seance in June 1928, Walter described an ambiguous Indigenous spirit as disrupting the project at hand: "Upset all my work! Upset all my work! ... My work! He did! That Indian devil!" When asked by T. Glen Hamilton to explain, Walter replied that a troublesome spirit he referred to as "Scarface" had "followed my friend Black Hawk" and "broke the chords" that Walter used to ring a bell.[145] In September the riotous spirit appeared again with Black Hawk. Described by Mary Marshall as "very dark" and "terrible looking," with a face "scarred and marked," he forebodingly stood beside Black Hawk throughout the seance.[146] Although Scarface received the blame for much of the upheaval, whereas Black Hawk was fairly compliant, distinguishing between the two proved difficult. A few nights later, Black Hawk appeared once more, this time alone, or so it seemed. His face, the medium declared, appeared "all scarred and marked" – his identity apparently melding with the "very dark Indian" of the night before to signal an equally menacing presence.[147]

Yet, despite this performance of the primitive, which was closely associated with violence and savagery, investigators and sitters of the Winnipeg seances consistently envisioned a submissive incarnation of Black Hawk who reflected the consequences of First Nation pacification rather than resistance.[148] From 1928 to 1929 Black Hawk refrained from taking his own initiative and willingly catered to the demands of Walter. When he made his presence known through the medium on December 23, 1928, Black Hawk relayed Walter's message and then referred to his own role as "just keeping guard ... I take my instructions from him."[149] In October 1929 Black Hawk informed the sitters of his even greater willingness to serve the interests of both Walter and the sitters: "I am here to do [Walter's] bidding. I am the messenger of anyone who will use me."[150]

Much like the amorphous White Bear, Black Hawk expressed his intent of "helping" the sitters in their investigations, while claiming his presumed "sins" for his lack of conformity to white-settler society. Although an instrumental conduit between the sitters and the otherworld, he claimed that he still needed to strive to meet the standards of holiness that whites allegedly possessed. Asked about Christ, he replied, "Me no get there to Christ yet. Much long way. Me must be good first; me no see him yet."[151]

The "work" Black Hawk engaged in with "Pale Face," or Walter, appeared to play a central role in his obtaining salvation. "Me work (hand in air). Me great work yet," Black Hawk claimed on several occasions.[152] Historian David R. Roediger has identified how colonial justifications of dispossession and violence centred on the conception of Indigenous peoples' failure to subdue and domesticate the land through agricultural labour. Such beliefs helped forge a representation of not only Indigenous peoples but also the "white worker." As Roediger argues, "Settlers, whether or not they worked harder or more steadily than Native Americans, came to consider themselves 'hard-working whites' in counterpoint to their imagination of Indian styles of life."[153] As a way of making amends for his deficiencies, the spirit guide Black Hawk worked toward the furtherance of paranormal revelations to the earthly plane in order to one day gain entry to the land of the "many, many great white spirits."[154] Whereas he admitted his own supposed guilt, he absolved the white sitters of theirs. Making no mention of his previous oppression or the violence of white colonists, Black Hawk instead emphasized the passage of time and the beauty of the land in which he now resided: "Many moons have gone. I know no years. I know no time. All summertime – beautiful, beautiful land!' ... all brothers in the spirit land ... all live in unity with one another."[155] Black Hawk conformed to the colonial ideas of the white sitters. In some regards, he enacted the historical pacification of Indigenous peoples, and at other moments, he expressed an unruly nature of which white investigators needed to be wary. The racist thinking of white sitters emblazoned on the spirit world strikingly revealed that psychical researchers did not exist beyond culture but were fully ingrained within it.

Expressing interwar anxieties that surrounded their experimental rooms, the authority of psychical researchers was challenged not only by unmanageable mediums and ghosts but also by the inability of researchers to be the so-called detached scientists they claimed to be.

***

Mediums, and the spirits that emerged from their allegedly submissive bodies, rested at the centre of psychical science, demonstrating profound paranormal power. Yet both mediums and spirits could also be the most unpredictable element of the seance. Psychical scientists endeavoured to categorize and pacify them in order to produce reliable results. Female mediums who defied sexual norms were critiqued and discredited. Male mediums, who threatened to disrupt gendered categories of mediumship, needed to be carefully managed. Working-class spirits required a reassertion of class hierarchies. Indigenous spirits who wreaked havoc during delicate paranormal experiments had to be subdued.

Yet mediums and spirits continued to challenge the power psychical researchers held within the seance room. Illustrating the extent to which interwar psychical science remained culturally entangled through transgressive performances, mediums and spirits opened up new possibilities by which empirical authority could be subverted and remade. Psychical researchers attempted to present their endeavours as fully in line with the scientific project of their era. Their investigations, however, aligned with the interwar world in other ways as well. The unstable subjects with whom they worked embodied circulating conceptions of gender, class, and race that ultimately undermined their investigators' claims to empirical authority. In the face of these subversive elements, psychical researchers continued to build their science of ghosts. The seance sensorium itself needed to be overhauled in order to reflect the ever-regimented standards of psychical science.

3
A Touch of the Uncanny: Sensing a Material Otherworld

> We have actually touched the borderland, where Matter and Force seem to merge into one another, the shadowy world between the Known and the Unknown.
> – William Crookes , "Crookes 'Radiant Matter'"[1]

For all its spiritual characteristics, the seance of the interwar period was strikingly physical. Indeed, it would seem that one came to know the workings of the spirits through not merely a supernatural sixth sense but also smells, touches, tastes, sounds, and sights. Slaps, feathered caresses, and inexplicable raps combined with candied tastes within the seance. Racialized spirits filled the room with perfumes and languages that evoked Western imaginings of Eastern knowledge.

Beginning with Marshall McLuhan, many scholars have argued that the increasing emphasis on the visual in the twentieth century had the consequence of diminishing other perceptions significantly.[2] This chapter and the next in part demonstrate just this shift in emphasis. What also becomes clear, however, is that the interwar seance represents one arena in which this alleged progression to the visual was not so straightforward or unproblematic, contributing to a body of work that complicates McLuhan's thesis.[3] Indeed, the multiple and varied sensual encounters of the seance supported the conviction that paranormal investigations could be remade into a modern and calculable science, compelling researchers to devise methods of measuring and classifying their sensory experiences within the seance. As spirits evoked gendered and racial bodies, psychical scientists insisted upon their own empirical perception. Investigators recategorized the senses of the seance room to reflect what they conceived of as standards necessary for the effective quantification of the preternatural. Coding their perceptual methods as objective, masculine, and hence distant from the body, psychical scientists established their experience of the paranormal as superior to what they regarded as more bodily and feminine sensory modes. Privileging hearing and sight for their disembodied and more objective qualities, investigators struggled with the limits of their unaided perceptions. Confronted with transgressive and frail embodiment on all sides, psychical researchers indeed found the seance to be truly haunted – by the spectre of corporeality.

The Seance Sensorium: Perfumes and Sweeties, Voices and Slaps

Psychical researchers employed their perceptions to interrogate the spirits who seemed ready to meet such demands. In fact, paranormal enthusiasts represented the seance more often as a space of heightened sensations than as a spiritualized diminishing of them – the psychical forces that emerged seeming "so material" that even the spirits could not always distinguish between the physical and metaphysical.[4] Otherwise invisible natural phenomena such as sound, light, and thought waves retained a decisively "tangible" quality for those beyond the grave, who arguably seemed to exist on a plane that was more material than that of the earthly seance participants.[5]

The senses played a pivotal role in the seance – a sitting typically overflowing with rousing perceptions from across the sensorial spectrum. In 1930 Jenny O'Hara Pincock wrote of a lively evening that was marked by fervent singing, wafting perfumes, supernormal breezes, and sprinkling water. Recalling the voice of her dead husband requesting the song "Joy to the World," she remarked, "as the song ascended our hearts were lightened. Exquisite perfume broke upon us. Breezes blew softly past our faces, the final benediction of water was lightly sprinkled on each bowed head."[6]

Child ghosts were known to demonstrate their telekinetic and taste-inducing capabilities during the O'Hara Pincock sittings by handfeeding seance participants candies and cookies. The spirit child of O'Hara Pincock, known as "Jane," declared one evening in 1931, "I have a candy ... I'll give you one now ... I'll put it in your mouth." Upon very precisely placing it on one sitter's tongue, she proudly noted, "You couldn't do that in the dark."[7] On the same night, Jane promised to repeat her impressive performance, except with cookie pieces. Her description of how she would prepare the cookie for consumption, however, demonstrated a desire to produce a very different taste sensation. Jane declared she would give each sitter a piece but only after she had "stepped on it, spit on it and rubbed some pussy hairs in it."[8] Even so, the sitters obediently opened their mouths as the spirit deftly supplied them with the aforementioned cookie pieces.[9] In this act, Jane supplied evidence of her presence, signalling her persistent, life-filled connection with the material.

Richly scented odours also wafted through several seances of this period, typically likened to floral aromas or musky colognes. Described as "treats from the spirit world" by discarnate intelligences, "beautiful perfumes of flowers" or some other distinct scent regularly filled the seance room.[10] Depending on the seance in which they manifested, mysterious odours varied from merely "faint" to intense and thick.[11] Uncanny smells were not isolated instances but accompanied the discarnate voices while they spoke, sang, and generally interacted with the sitters. The seance record noted how, while the sitters were singing along with the ghostly voices of O'Hara Pincock's deceased husband and her spirit child, Jane, "an exquisite perfume-laden breeze broke upon us."[12]

Touch also played a dominant role within the seance – the spirits eagerly caressing, prodding, poking, and tussling with the sitters. O'Hara Pincock recorded how sitters "were patted on the face, cheeks and hands" and "many times felt hands lovingly caressing" them. They were even "banged in the stomach" at other moments.[13] Certain touches seemed to be extremely evidential. The ghost of O'Hara Pincock's father commonly made himself known by a "rat-a-tat" on the head of O'Hara Pincock and others "very much like he used to tap with his fingers while on earth."[14] Of a ghostly finger, Pincock similarly reported, "It felt like Papa's – large and solid and it certainly wasn't a human one, for it had a most peculiar putty-like, cold feeling."[15]

Perhaps one of the most valued perceptions in terms of how it was thought to relay evidence of the otherworld was the sense of hearing. In fact, it was entirely common for the initial form of paranormal communication to occur through a series of table raps. Stemming from a long tradition of nineteenth-century seances, table raps played an important role in indicating the presence of supernatural beings capable of spiritual telegraphy.[16] To encourage such spirit communication, sitters used music, which only added to the auditory quality of the seance. Singing hymns or playing a phonograph reportedly produced the appropriate amount of "vibrations" necessary for materialization. Some argued that singing or playing music relaxed the psychic, especially if it was a "certain music that they liked," and thus succeeded in "minimizing the medium's natural inhibitions to the trance state."[17] The type of music shifted depending on the seance and the whim of the spirit who dominated at a given time. Hymn singing almost always took place in the sittings over which O'Hara Pincock presided at the New York spiritualist retreat of Lily Dale and in St. Catharines, Ontario, her piano providing the music that brought the spirits forth. Pincock recorded one 1928 seance of this nature, which she described as "pregnant with evidence and marvellous manifestations." On September 9 a ghost engaged the sitters with spiritual violin music of his own. The spirit could do so only after the piano had provided the necessary "vibrations," from which he "secured enough strength to materialize for the performance that followed."[18] Such hymns, the spirit Dr. Anderson argued, provided the appropriate sort of vibrations, in contrast to jazz, for instance, which apparently drew "entities whom we might not be able to control ... guests as carnal minded as those of earth, dancing and carousing."[19]

The Walter spirit of both Boston and Winnipeg also seemed to rely heavily upon music but of an utterly different tenor. This notoriously rowdy ghost demanded popular tunes to be sung, ranging from "Tramp, Tramp, Tramp" and "Golden Slippers" to "Clementine" and "Jingle Bells."[20] To add to the chorus, Walter commonly whistled along "in time and in tune" with the music. He proved so adept at this feat, in fact, that his characteristic whistle soon became a standard feature of the mediumship. Opinion varied as to the musical quality of Walter's whistle, but many investigators described it as "skillful."[21] Despite the spirits' resistance to "canned music," both the Winnipeg and Boston seances frequently used the phonograph, or "Victrola."[22] Since the Boston Walter was extremely fond of the song "Souvenir," the tune regularly played during Margery's seances. From January 30, 1924, onward, Margery's investigators continuously played a "saxophone solo" of Walter's beloved song during sittings in order to encourage phenomena to manifest.[23]

Of all the supernatural and earthly sounds, however, the voices of the discarnate beings were prized by those who explored the paranormal. In T. Glen Hamilton's Winnipeg experiments, the voice commonly emerged directly from the vocal chords of whomever the spirits chose to control, a phenomenon investigators referred to as "automatic voice." At times, however, the spirits allegedly chose to express themselves through independent disembodied voices completely external to the medium, or "direct voice," as researchers labelled it.[24] Mediums typically demonstrated only one of these manifestations, but some proved able to do both. Such was the case with Mary Marshall in the Hamilton seances, as the automatic voice occasionally gave over to the direct voice.[25] Margery used automatic voice, but her mediumship, as well as that of William Cartheuser, also used direct voice, in which the discarnate being spoke from some identified point in the room apart from the medium or, in Cartheuser's case, through an aluminum trumpet.

The seance trumpet of the twentieth century resembled much older sound-augmenting devices from as far back as the sixteenth and seventeenth centuries, particularly the ear trumpet, or "otacousticon." Scholar Leigh Eric Schmidt charts the development of these technologies alongside occultist traditions, marking "the interplay between such magical traditions and the acoustic art."[26] The ear trumpet was not only used to aid those with hearing loss. The device was also believed to carry sound across impossible distances or even from one plane of existence to another and was thus said to "bring down the spirits into material form and to allow them to speak."[27]

By the twentieth century, the trumpet had apparently not lost its appeal. According to spiritualist Anna Louise Fletcher, the discarnate voice speaking through the humble aluminum cone far surpassed any other paranormal phenomena, as it could reveal the authenticity of both the particular ghost in question and the medium who encouraged the spirit's emergence. Fletcher wrote to Cartheuser in 1927, shortly after one of his discouraging encounters with the American Society for Psychical Research (ASPR). She expressed how "distressed" she was "to know that you have suffered at the hands of so-called scientific researchers – a name they have not in any way merited by the methods they employ in searching for the soul." Fletcher argued that, to judge from a sitting she had held with him at Lily Dale, even if Cartheuser "had tried to manipulate the trumpet with your hands and feet, you could not have produced my father's voice, his characteristic way of expressing himself – nor could you have known the least thing about him." In Fletcher's mind, the efforts of psychical researchers to establish the credibility of mediumistic gifts focused too much on visual and tactual phenomena. In contrast, she remained convinced that the voices that emerged from the trumpet formulated "the evidence on which to pin our faith." Distancing both herself and Cartheuser from such harsh methods, she comforted him with the assurance that "I know your work is perfectly honest and trustworthy. I wish I could say the same for some of the 'researchers.'"[28]

Despite their different approaches, the psychical researchers that Fletcher so harshly criticized also placed a great amount of importance on the voice. Not only did they appraise the messages conveyed, discerning certain details as evidence of the spirit's authenticity, but they also took note of the specific vocal tone that differentiated the disembodied speaker from the medium. Cartheuser's mediumship frequently led to a plethora of unique voices. Dr. Anderson's voice was purportedly "entirely different" from those of the other spirits who spoke during his seances.[29] Moreover, O'Hara Pincock claimed to recognize the idiosyncrasies of her dead husband's voice on numerous occasions during sittings with Cartheuser, including "a familiar warble known well by his loved ones while he was on earth."[30] Even the skeptical Henry Clay McComas admired the voices that emerged from Cartheuser's mediumship. When studying the voice of Cartheuser's little girl spirit named "Elsie," McComas admitted, "It did not seem possible that Cartheuser could stretch his vocal chords to reach such high tones."[31] McComas claimed to identify how Cartheuser created this effect through bodily rather than spiritual skill. He nevertheless confessed that Cartheuser's "capacity to bring such tension of the cords [so] that a man's voice appears to be that of a child is most exceptional."[32]

The spirit Walter, almost without exception, manifested through female mediums. Investigators' fervent avowal of Walter's manly characteristics made it crucial that his voice be recognized as appropriately masculine in contrast to the voices of his womanly psychics. In several instances, the voice of Walter lived up to the task. In the Winnipeg seances, observers reported that Walter's discarnate voice, emerging either directly from or in the vicinity of Mary Marshall, had "a manlike quality ... husky, but vigorous."[33] Similarly, in the Boston seances, the Walter voice was reported as "very husky; distinctly masculine," and "beyond Margery's normal range."[34] Nevertheless, not everyone was convinced. Upon sitting with Margery in 1925, B. Wentworth Emmons declared that after listening to the whispering voice of Walter, he determined that it could be understood only as that of Margery, however disguised. "Especial attention was made to compare the modulation and articulation of the whispered voice to Mrs. Crandon's [Margery's] voice and they seemed to be identical, the phrasing and intonations were very similar."[35] Joseph B. Rhine meanwhile, skeptical as he was of mediumship in its entirety, very bluntly described the Walter voice the following year as "resembling [that] of a woman whispering hoarsely to imitate a man's voice."[36]

Evoking the perceptions of the sitters, spirits indicated their material, even embodied, presence in the seance room. Moreover, such sensory experiences demonstrated how entangled the supernatural and the natural worlds remained. Anthropologist David Howes has argued that the senses "are imbued with meaning and carefully hierarchized and regulated so as to express and enforce the social and cosmic order."[37] Rather than being simply a series of stimulations, the senses became a means by which investigators and sitters could "experience ... gender, colonialism and material culture."[38] As products of culture and history, the touches, smells, and sounds of the seance reflected conceptions of race and gender that circulated in the interwar context.

Sensations of Orientalism: Spirits, Race, and Accessing the Other

Racialized spirits associated with Western representations of the Orient added to the sensual encounter of the interwar seance room. By arousing the perceptions of the mediums and sitters, these spirits shaped the seance conditions while providing rich sensory evidence. Evoking pervasive impressions of the East, infused with Western ideas of racial difference, these ghosts reinforced conceptions of Orientalism through which an exotic other was sensually created. By means of this "radical realism," as scholar Edward Said describes, the Orient was placed in opposition to the white sitters through a fantastical set of loose associations that were at once imagined and perceived.[39]

Spirits evoking the Orient initially presented themselves in the Boston seances in 1923. On the evenings of June 23 and June 29, the medium Margery scrawled a few phrases in "Anglicized Chinese."[40] Automatic writing of this nature continued throughout the 1920s and 1930s at Margery's seances, signalling the presence of spirits referred to as "Kung-fu-tze" and "Tao." The researchers took note of how these spirits created a "distinctly oriental Chinese atmosphere" in the psychical laboratory. For successful experimentation, the sitters needed to perform alongside the spirits in order to create a racially defined environment within the seance room. "We must be oriental tonight," Walter announced at the beginning of a seance in March 1928.[41]

The presence of these foreign ghosts was not welcomed by all, particularly Walter, who complained that he "doesn't much relish the constant presence of all these Chinks," who apparently persisted in "high-hatting" him. During one spring seance in 1928, Walter again voiced his reluctance to participate in the interracial experiment: "I don't mix with Chinese myself, but it isn't my choice." These spirits continued to appear nonetheless, providing several pieces of sensory evidence, or as investigator Mark Richardson described, "an oriental flavour."[42]

Spirit guides identified as "Chinese" commonly announced their presence in the seances through auditory means, either by distinct table raps, music, or direct voice. They introduced themselves through "many repetitions of a series of light, fast raps which regular sitters have learned to associate with the purported presence of the oriental controls."[43] On one evening in Boston, investigators deciphered their presence by the sound of a "flute." This was then followed by the recital of a proverb attributed to Confucius, a "greeting customary in China," and various other "ceremonial phrases." Only vaguely referring to what the spirits said, the investigators reinforced the idea that the words of the foreign ghosts were significant only insofar as they evoked an amorphous East, not because of the precise message they conveyed.[44]

Characterized by broken or mispronounced English with exotic auditory embellishments, the communications of Oriental spirit guides conformed to the white, middle-class conceptions of racial difference. In another seance of 1928, this time in Winnipeg, an occasional medium known as "Mrs. Hurling" communed with a spirit who was understood to be Japanese. He made his difference from his white audience audible through a language that was "very strange and difficult to understand."[45] Similarly, during a 1929 seance in St. Catharines, the spirit known as "Oolong" shook a "bracelet with bangles on it ... in time to the music," while he entertained the sitters by "talking a funny language which he said was Chinese."[46]

Other racial types associated with the East also made themselves known. During one Winnipeg experiment in February 1928, the entranced Mary Marshall spoke in Hindustani to one of her investigators, W.B. Cooper. This moment held much significance for Cooper, who had grown up in India and was familiar with the language.[47] Eastern spirits tended to stimulate perceptions of smell as well as hearing. In a seance on March 4, 1928, Marshall once more went into trance, shivered, and sighed. Suddenly, a "strong odour" filled the air, which all agreed smelled "like cologne." Upon regaining consciousness, Marshall described a vision that made the racial overtones of the olfactory sensation clear to everyone: she was "on a boat with a dark-skinned man. He wore a turban and robes. I think he was a priest. He took me to a large temple. It had twelve pillars, and oh! The beautiful lights! ... All praying, kneeling in prayer. Their robes were gauzy white."[48]

Historian Mark M. Smith has argued that race is a construct that frequently draws upon all the senses – not just vision but also hearing, touch, taste, and smell. Together, these "proximate, nonvisual senses ... invent 'modern' racial stereotypes," which in turn highlight "the historically conditioned, visceral, emotional aspect of racial construction and racism."[49] So too did the seance participants attest to the presence of something racially divergent from their white selves through hearing and smell. These exotic ghosts, made apparent through the sitters' sensory imaginations, domesticated the Orient to make the East consumable and easily controlled.

Eagerness to meet such exotic ghosts in the seance coincided with international racist policies aimed exclusively at those of East Asian descent. So harsh were these policies that historians have referred to this period as the "Exclusion Era," which was marked by immigration restrictions, discriminatory laws, limited to no opportunities for enfranchisement, and mob violence.[50] Scholars have connected such acts of exclusion to a broader global trend of the late nineteenth and early twentieth centuries. In the interests of defending the alleged racial purity of "white men's countries," a number of policy makers, government officials, and activists lobbied for and employed various strategies for prohibiting immigration, while advocating deportation and segregation.[51] Arguments for the need to assimilate Asian immigrants coupled with fears that democratic traditions would be undermined if the races intermingled.[52]

Oriental spirits reinforced the sense of distance white sitters felt between themselves and a racial other. Through their broken language and sensual enactments, Eastern apparitions affirmed presumptions of white superiority, while fulfilling the desires of white seance participants for the exotic and the extraordinary. Much like scholar Homi Bhabha has argued, such "productive ambivalence" forged "that 'otherness' which is at once an object of desire and derision, an articulation of difference contained within the fantasy of origin and identity."[53]

Allusions to an amorphous East in the seance reflected the belief that somehow these spirits granted white seance participants an added capacity to transcend the earthly. Evoking a mysticism that was at once interior and individualized, the spirits did not challenge assumed hierarchies between the West and the Eastern other but merely provided an easily digestible form of personal spirituality that catered to the needs of the white sitters.[54] As the spirit Dr. Anderson claimed in a St. Catharines seance in 1932, Hindu spirits standing just outside the circle acted as each seance participant's "own Hindu." The sole purpose of these Eastern spirits, Dr. Anderson contended, was to "develop" each sitter's capacity for "healing and listen to your call for aid."[55]

The manner in which these ghosts stimulated the senses of both sitter and medium revealed that such a heightened form of mysticism could be experienced through proximity to the body. Eastern religion was commonly characterized as relying upon the imagination, the "mistress of the senses," which titillated bodily perceptions as well as the mind.[56] Exotic odours streaming from the medium emphasized the more carnal perceptions that these ghosts aroused. Coming "in waves," musky perfumes associated with the Orient poured from the psychic's body, specifically her breasts, while she became "greatly excited."[57] The way that the Eastern spirits possessed the medium suggested a sensual encounter that completely engulfed and transformed her.

To symbolize her bodily possession by an exotic other, Margery began to occasionally wear a kimono during experiments.[58] Mari Yoshihara has argued that white women participated in producing the effects of Orientalism through consumption of Asian goods and theatrical cross-dressing. Mimicry of an Eastern otherness enabled them to transgress regimented gender and sexual roles. With the rise of the "new woman" and "modern girl," such mimicry allowed some latitude for white women to exteriorize shifting gendered and sexual mores in the first half of the twentieth century.[59] White women dressing up and performing stereotypical conceptions of Oriental womanhood came with fewer complications than one might think. The East was not only effeminized in opposition to a Western masculinity but was sexualized as well.[60]

Margery's donning of the kimono marked her self-conflation with the sensual Orient. Through her, the spirits produced a body of a sexualized other that could be conveniently prodded and scrutinized. Investigators did not hesitate to take full advantage. The psychical researcher seized upon these incarnations of the medium's body as an ideal means by which, like European Orientalists, he "could penetrate, he could wrestle with, he could give shape and meaning to the great Asiatic mystery."[61] Yet, rather than emphasizing the carnal "fantasy of the senses" that these encounters characteristically evoked, the investigators needed to render the experiences of the seance empirical and rational.[62] Psychical researchers reoriented the seance sensorium to reflect their scientific, masculine ambitions by presenting their faculties of perception as superior to those of mediums and other seance participants.

Feeling Scientific: Odourless Bodies and Controls

The difficulties that faced an investigator when entering the seance room were numerous – the most distracting of all, perhaps, being the sudden caresses, slaps, and lingering touches by what psychologist and psychical researcher Henry Clay McComas described as a "clammy finger" or "a trumpet ... on the forehead." Such peculiar sensations came "so unexpectedly," McComas complained, that one could "hardly describe the occurrence at all."[63] "Subjective sensations" such as these often endangered rather than aided an investigation of the paranormal. The touches, sounds, smells, and tastes of the spirits might very well divert researchers from their primary duty of collecting and discerning credible evidence.[64] Psychical scientists such as Eric J. Dingwall differentiated themselves from "idle seekers of sensation" by arguing for the necessity to employ their own acute perceptions when studying the paranormal in order to counter possibly spurious impressions that the spirits induced.[65] Smell was periodically used, but the sense became suspect as soon as any particular scent was detected. Touch was considered to be wildly misleading and promiscuous. Even still, an investigator could use the perception as an aid to observation, as well as a means to control the medium's body, but only if he "kept his composure."[66]

Floral and richly aromatic scents frequently made their way into the seance room, a phenomenon that psychical scientists found to be problematic for the most part.[67] As scholars Constance Classen, David Howes, and Anthony Synnott have argued, from the nineteenth century onward the sense of smell was increasingly represented as feminine and, correspondingly, as ill-suited for the objective and scientific gathering of information from the surrounding world. In opposition to vision, which was a perception aligned with manly reason, smell came to be "considered the sense of intuition and sentiment, of home-making and seduction, all of which were associated with women."[68] Floral or sweet odours were especially feminized.[69] Perhaps for this reason, psychical investigators reported instances of perfume and floral scents as interesting but inherently imprecise. J. Malcolm Bird described a seance in which sitters experienced the feeling of tufts of hair or fur "heavily scented, held under the nose." He complained that the "odour went unidentified, unless a combination of rose, violet or lily of the valley identifies it."[70]

Investigators consequently moved their attention away from olfactory-inducing experiments to focus upon the smell of materializations. More skeptical individuals, such as Walter Franklin Prince, alleged that paranormal manifestations typically "smelt of rubber," indicating his suspicion that they were manmade artifices rather than spiritual beings.[71] In contrast, those who were more convinced of the paranormal phenomena they perceived claimed that psychic bodies, or at least genuine ones, had no smell whatsoever. During a 1925 seance in Boston, one of the investigators reportedly brought "his nose very close" to the mysterious paranormal mass, only to describe that it apparently had "no odour of any description."[72] A year later, Unitarian Church minister John Haynes Holmes reported that in addition to it being "cool" and resembling "a fairly solid jelly," the materialized mass he handled had "no odour" that he could detect.[73] Such reports stood in contrast to the floral wafts of apparently less verifiable phenomena – the lack of odour validating the phenomenon's authenticity. Suggestions made my Dingwall, William McDougall, and others that the materialization resembled some sort of contraption made of animal membrane were contradicted by its lack of odour, organic or otherwise.[74] Even more significantly, this absence of odour redirected such apparitions from the realm of the superstitious to that of the scientific. As Classen argues, the olfactory, once closely aligned with the sacred, came to be regarded as inherently pathological, especially with the formulation of what twentieth-century psychologists referred to as "olfactory hallucination."[75] Classen describes how, correlated with "sexual and mental disorders," it was associated with certain "forms of religious insanity."[76] To be odourless was to be scientific and rational, attributes that psychical scientists desired for themselves as well as their objects of study.

Much like smell, touch was also experienced and employed by investigators. In the seance room, touch could mean many things, ranging from ghostly caresses to firm grips and meticulous inspections to the feel of ropes, wires, and tape pressed against the psychic's skin. Psychical researchers used touch for the purposes of general observation of phenomena as well as restraint or inspection of the medium's body. Due to the dark conditions of the seance room, one could not rely solely on sight and hearing, as Hereward Carrington admitted. Rather, one needed to rely on any "tactile observations as one could make."[77] Consequently, investigators noted their physical encounters with paranormal phenomena as having a specific feel and texture. Describing mysterious materializations as being "soft," "very cold," and "wet," psychical investigators attested to their materiality, which apparently left a lasting impression. As Bird testified, "After one's skin had been touched by it, the cold, moist feeling persisted for several minutes."[78]

Touch, like smell, could nevertheless be misleading due to its intimate proximity to the body. Both Classen and philosopher Carolyn Korsmeyer argue that touch has been associated with "irrationalism and primitiveness," making it a lower sense aligned with the feminine. Although perhaps not as disdained as smell, touch could nevertheless "focus attention on the conditions of one's body," which in turn led one to be more subjective than an objective and logical view of the world allowed for.[79] As a result, psychical researchers attempted to manage tactile perceptions in the seance, representing themselves as the only individuals in the room credible enough to employ this potentially wayward perception. Not everyone easily assented to these prescribed limitations. Some mediums spoke out in opposition to such protocol. As Cartheuser lamented, psychical researchers' prohibition of touch by anyone but themselves impeded spirits, who "inspir[e] you or impresses you to touch an article with your own hands." If psychical researchers detected such illicit touching of objects, they quickly condemned the act, arguing that genuine psychics "are not supposed to touch it. They are supposed to send it into the air in some mysterious way." Cartheuser argued, however, that the spirits sometimes needed the medium to lift "the article to a certain degree until ectoplasmic force can reach it."[80]

Despite such resistance, psychical scientists insisted that the legitimacy of their pursuits could be maintained only if spurious sensations such as touch were employed by no one other than the investigator in authority. Researchers applied their own sense of touch in the seance, usually in the form of what they called controls. The necessity for scientific controls in the seance remained in keeping with other twentieth-century scientific enterprises, which aimed to eradicate external forces that would skew experimental results. In psychical science, however, controls often translated very literally to mean restraint and scrutiny of the psychic's body, the biological centre through which the spirits manifested. Such precautions aimed to limit the number of factors that were not supernormal and thereby to "make fraud impossible, so that anything occurring is stamped at once as genuine."[81] Psychical researchers typically viewed a medium's apprehension toward controls and testing with suspicion, insisting that a medium could be verified only if he or she submitted to the full rigour of scientific examination. When reflecting on a series of investigations of Cartheuser, in which the medium refused to comply with any of his demands, Carrington complained, "It is of course absurd that a medium should be permitted to dictate the conditions prevailing at his sittings."[82] The imposition of controls was the only thing that could secure reliable findings.[83]

Lorraine Daston and Peter Galison have asserted that "first and foremost, objectivity is the suppression of some aspect of the self, the countering of subjectivity."[84] Objectivity was conceived of as an effort to "negate subjectivity" in that "the emergence of objectivity must tally with the emergence of a certain kind of wilful self, one perceived as endangering scientific knowledge."[85] Within the walls of the psychical laboratory, this countering of subjectivity took on a material and sensory significance. Whereas the psychical scientist engineered his identity to encapsulate objective perception, he cast the medium as the embodiment of irrational and subjective sensory modes.

The medium and the sensual paranormal phenomena that emerged in the seance thus came to symbolize all that the psychical scientist needed to regulate and master. Several investigators insisted that the medium be clothed in a standardized set of clothing and restrained in some fashion throughout the seance. As T. Glen Hamilton advised a prospective investigator, "I would suggest that if you have creatures attempting to put on materialization stunts, that they be compelled to wear black stockings and black shoes ... that every precaution be taken in the way of holding their hands during performances."[86] True to his word, Hamilton had women assistants systematically strip, inspect, bath, and re-dress the psychic in a standard seance garb for each experiment.[87] The medium's hands and occasionally legs were restrained to ensure that accusations of fraud remained unfounded.[88]

Margery first experienced such controls in December 1923 during her trip to Europe. Investigators insisted that Margery not only have her hands held but also have her legs bound, while two men, one on either side, rested their heads on each of her shoulders. "The control was the most severe [to which] Margery had up to this time been subjected," Bird related.[89] When she returned to the United States, her investigators continued this practice of restraint, becoming increasingly aggressive with each experiment. In one 1924 seance, Harvard professor William McDougall held "both of Margery's legs in his usual vice-like control," and Abraham A. Roback's inspection of the medium's body was described as "roving."[90] In March 1924 McComas reportedly "held Margery's legs between his so long and so tightly that his sensations were numbed."[91]

In certain circumstances, the violence of such restraint revealed itself on the medium's body. Although Hamilton in Winnipeg safeguarded the general comfort and well-being of his mediums, several of Margery's more cynical investigators were less concerned with the health of the medium. As control of her body became more intense, Margery reported swelling and discomfort. In 1926, following an evening of rigid control by her investigators, Margery stated, "I'm in awful pain," to which the spirit Walter replied, "Nobody gives a damn if you are."[92] As historian Linda Gordon first illustrated, the physical abuse of women in this period may have been officially illegal, but it was still largely tolerated and normalized.[93] Moreover, because this sort of manhandling, so to speak, coincided with stereotypically masculine modes of touch and aggression, it fell more easily in line with the project of psychical science.[94]

Mechanizing Touch: Automatic Controls and Systemized Inspections

However, tactual restraint of the medium's body, or "personal control," was not free of its own problems, as investigators distrusted even one another's ability to manage the spurious perception of touch. Consequently, mechanical or automatic controls were introduced into the seance to eliminate the reliance upon subjective sensations and opinions. As J. Malcolm Bird put it while in the midst of the investigation of Margery sponsored by *Scientific American* in 1924,

> Shall one rely for validation of the phenomena upon the traditional control of the medium, plus systematic exclusion of the opportunities for confederacy? Or shall one seek, with the aid of modern scientific apparatus, to put the investigation on the basis of machine-made precision, which shall eliminate the bugaboo of the accomplice, and afford us an automatic guarantee that anything that happens in the seance room is necessarily genuine?[95]

The *Scientific American* committee members used personal control for the first official sittings with Margery, but increasing tension over allegations of fraud led to the introduction of new forms of restraint. Physicist Daniel Frost Comstock urged that mechanical controls be employed. Their use would ensure that the subjective bodies of the investigators or anyone else did not get in the way of making conclusive assessments. Comstock's convictions led to the use of mechanical touch within the seance. Rather than various investigators holding the medium's extremities, she would – at least for some of the experiments – be tied down using ropes, tape, wires, or in select cases, even more elaborate methods.[96]

The first stage in imposing this automatic sense of touch involved the committee asking Harry Houdini to "manufacture [a] control apparatus which would leave Margery comfortable and which under his guarantee, would prevent fraud." Although initially assumed by Le Roi Crandon to mean something along the lines of handcuffs or ropes, Houdini instead built a large wooden cage in which Margery would be sealed. In the interests of making it secure, padlocks were added so that the medium could be freed only by the one who kept the keys.[97] Her head and hands, coming out from strategic openings on the top and sides, were then held by her investigators. In such a disciplinary realm, one might suppose that absolute control of the medium had finally been achieved.

Accusations of fraud against Margery nonetheless continued even in this context, leading to the application of several more mechanical restraints. Although not always going to the extreme of Houdini's cage, Margery regularly succumbed to being tied down in some fashion, despite her initial objections to such tactics.[98] During a 1924 experiment, McDougall insisted that the medium be bound and tied to him by a loop of rope.[99] The following year, investigators began binding Margery's wrists and ankles with strong picture wire. Even after investigators covered the wires with rubber tubing, the ties still cut into the medium's skin, leaving her sore for days afterward.[100]

In 1926 surgical tape and plaster were used to restrain the medium, in addition to the previous methods. Investigators affixed tape to Margery's bloomers, stockings, and slippers, each strip marked with blue pencil that extended across to her skin. Her wrists and ankles were bound with wire using a combination of square and surgical knots, which were then tied to the cabinet and secured with a "standard lead seal." Finally, "surgeon's plaster was passed round the lowest two inches of the thighs, around the knees and down the legs two inches, binding tightly thus, the knees together."[101]

Margery's investigators justified the violence of these mechanical controls by insisting that they would protect the medium from unreliable tactual restraints and thereby save her from unfounded accusations of fraud. Whereas someone holding the medium's hands could possibly lie or be tricked, physical bounds could do neither. Only mechanical controls, being distant from the subjective body, could establish "scientific procedure" within the seance, despite the pain they caused the medium.[102] Perhaps just as importantly, they guaranteed that the tactual faculties of the psychic would be effectively held at bay. Investigator William Button claimed that these forms of control, "if properly exercised," would make "the medium helpless as to any normal participation in the manipulation of objects."[103] Creating what Michel Foucault might call "docile bodies" through mechanizing touch, psychical scientists carefully regulated bodies in the seance – even their own – so that verifiable evidence of paranormal phenomena could be obtained.[104]

Yet what some viewed as satisfactory and even extreme, others saw as lenient, and arguments continued over whether these sorts of mechanical controls were stringent enough. Joseph B. Rhine expressed frustration at the fact that Margery's wire restraints measured three or four inches long, which according to him "clearly enabled her to draw her hands inside the cabinet, where, with her slender wrists allowing the rubber-insulated wire to slip further up, she doubtless could have a reaching radius of about ten inches, varying according to the tying."[105] As a consequence of such skepticism and suspicion, personal touch was reintroduced to the seance in order to provide an ever-increasing scrutiny of the psychic's body. In contrast to the spurious touches of the medium and spirits, however, this touch would be regimented and systematic.

The idea of thorough inspections initially emerged in the mid-1920s. Although Crandon refused to allow Margery to be examined internally, he eventually agreed to thorough external examinations. Margery herself very frankly introduced the first examination of her clothing and body when holding a solo sitting with Walter Franklin Prince in 1924. After Prince expressed concern during a sitting that she could potentially conceal something under her gown, Margery disrobed in front of him and demanded that he examine both herself and her garment there and then.[106]

This practice of inspection continued, yet it would no longer be a result of Margery's initiative but would be due to the insistence of her investigators. At the outset of Eric J. Dingwall's experiments with Margery from December 1924 to January 1925, he stipulated that Margery's body be meticulously examined prior to experiments. Others who joined him in his experiments, including William McDougall and Episcopalian minister Rev. Dr. Elwood Worcester, supported Dingwall in pressuring Margery to wear nothing more than a robe, stockings, and shoes to each sitting, as well as to submit to an intimate tactual examination of both her garments and body before and after the seance. At the beginning of the investigations, Margery "was roughly searched" by Worcester.[107] Such examinations continued for the entirety of their experiments.[108]

Scrutiny of Margery's body and clothing happened before and after official seances throughout the next several years, with varying degrees of thoroughness. In late 1925 Bird complained that these investigations needed to become much more systematic and standardized. His report put forward by the ASPR recommended that in preparation for a seance with Margery, "a formal, fixed routine for search of the psychic be agreed upon."[109] In apparent response to such calls for regularity, examinations grew increasingly rigorous and lengthy. Following a specified order and method, hands moved across Margery's body, clothing, and bindings before and after each sitting.[110] By 1931 the thoroughness of the inspection was made apparent by detailed and technical reports. R.J. Tillyard, a respected entomologist, solicited a medical doctor, Joseph J. Skirball, to accompany him during his sitting with Margery in early 1931. Skirball's duties involved conducting a physical examination of Margery before and after Tillyard's sitting with the medium. His report revealed minute details of skin texture and abrasions, most likely caused by her bonds from previous experiments. Skirball noted "an area of denuded epithelium together with subculicular haemorrhage one and three quarter inches by half inch on the anterior aspect of the left forearm extending to the point about two inches above the upper end of the taping."[111]

Mechanizing and systemizing touch became an increasingly important part of the psychical researcher's program. Whereas the medium's tactual faculties were negated altogether, the touch of the psychical scientist was either replaced with automatic controls or made more regimented and standardized. Despite the effort of investigators to manage touch throughout the process of frequently violent control and inspection, it continued to be viewed as less superior to more disembodied senses.

Heightened Perceptions: Disembodying the Psychical Senses

Psychical investigators insisted that the sensory impulses inherent to the seance needed to conform to their standards of disembodied objectivity. All was not equal in scientific value. Investigators argued that certain senses more effectively gathered objective information, especially those deemed more distant from the body. Despite frequently relying on the so-called lower perceptions of touch and smell, psychical researchers idealized the allegedly higher senses of hearing and vision, even as both met with significant impediments in the seance room.

Evelyn Fox Keller and Donna Haraway have identified how the claim to objective knowledge of the world rested upon the ability to distance oneself from the body, which only men could attain. Fox Keller describes this requirement of objectivity as the twentieth century's "conundrum" of the natural sciences, whose metaphors and practices defined "the scientific mind ... as both male and disembodied."[112] Following this logic, perceptions that hearkened to the body were maligned, whereas the more disembodied senses were prized. Touch, smell, and taste were "the 'bodily' senses" and thus almost unavoidably subjective. Vision and hearing were ascribed a higher status because they remained distant from the body and were therefore believed to be truly objective.[113] So-called bodily senses were coded female, whereas the heightened disembodied perceptions were deemed masculine.[114] Such a philosophical hierarchy of the senses demanded that one had to experience the world in a particular way for certain sensory practices to be counted as valuable.[115]

Psychical researchers used this sensual hierarchy for their own ends, prioritizing one type of perception over others in relaying evidence, while dictating the methods of collecting, recording, and interpreting sensory experiences. Hearing, deemed a "higher" and thus masculine sense, was believed to reflect objective, scientific standards. Much as Korsmeyer has argued, this perception remained closely aligned with a masculine ability to "transcend the body" for the purpose of collecting sensory information from a world that remained fundamentally "remote" from the observer.[116] The experimenters emphasized their accurate and precise detection of various voices and noises, whether supernormal or not, from around the room. Documenting and detailing the various auditory stimuli in depth, investigators presented their particular methods of aural perception as aligned with objective and, by implication, masculine principles. Hearing the voice of discarnate intelligences in the seance thus became less about listening to the actual messages conveyed in order to discern their specific metaphysical and personal significance and more about accurately pinpointing the precise origin of the voice in the room. Documenting the voice's development, investigators verified that it adhered to the scientific principles of the seance-turned-laboratory. Other sounds likewise became classified, measured, and standardized to reflect manly scientific ideals of precision and objectivity.

Investigators began to suspect that Margery's mediumship involved the phenomenon of direct voice in 1923. Starting as little more than a "soft formless blowing," the voice became "sibilant" over time, forming simple responses such as "yes" or "no." Gradually, its messages became more complex. Recorded as coming "from the cabinet and from other places about the table," as well as from "various parts of the room," the voice was listened to intently so as to verify its origin. Bird reported that Walter's voice typically "came too high up for Margery's mouth." At other moments, Walter's "whistling occurred with considerable freedom in and around the cabinet" or even emerged "most certainly from another direction than [Margery's], in red light."[117] Listening intently, Bird also noted a phenomenon he described as "a curious combined laugh," which consisted of the laughter of both Margery and Walter, coming "from a common point in space, and they gave the impression of being tangled up together, as though conceivably from a common physical organism." Bird's fellow investigator, Abraham A. Roback, concurred, claiming that "they were distinctly two sounds mingled, and not a single eccentric sound."[118]

On March 2, 1923, similar to Cartheuser's spirits, the Boston Walter began to speak through an aluminum trumpet, or "megaphone," "with great directness and force." As with Cartheuser's Dr. Anderson and Elsie, the trumpet moved freely about the room, the voice apparently speaking from multiple directions at any given moment. This phenomenon continued, but rather than dwelling on the meaning of the actual words spoken, Bird insisted that the chief duty of the psychical scientist was to detect the origin of the voice. He dismissed ventriloquism since it required some form of visual distraction that could not succeed in the dark conditions of the seance room. Bird then went on to present an anatomical explanation of the ear's accuracy in detecting the direction from which a particular sound came, leading him to argue that his own and his colleagues' acute sense of hearing could be relied upon to discern from where the voice spoke. Supporting his argument with detailed aural observations during a seance on May 19, 1923, Bird recorded how they meticulously counted the number of whistles and whispers, distinguishing them from other noises in the room. They timed Margery's respirations in relation to these sounds, which were carefully accounted for while investigators took turns restraining Margery's mouth with their hands.[119]

Bird and Roback were by no means alone in emphasizing their scientific analysis of the voice and its origins. Throughout B. Wentworth Emmons's records of experiments with Margery, he described his vast ability to discern and verify a compendium of noises. Remarking on the discarnate voice, he explained that he "distinctly heard whispering coming apparently from Mrs. Crandon [Margery]." His attempts to lean in closer to analyze the voice met with opposition from the spirit Walter, who informed Emmons that he should "assume a more comfortable position." Emmons nevertheless continued describing the voice in detail, noting its fairly "natural" quality and the "shrill notes" of Walter's whistle. He also recorded his detection of other, even mundane noises beyond the voice, including "taps or raps" against the table and the "occasional scraping and knock of the repeater upon the victrola" or "the striking of the clock." When Le Roi Crandon supplied the spirit Walter with a Spanish gourd, Emmons outlined the sound it emitted upon rolling about the table. This came accompanied by "delicate raps" and "three blows ... struck on the table near my right hand."[120]

T. Glen Hamilton in Winnipeg treated the occurring voices and sounds of his psychical laboratory in a similar manner, demonstrating his capabilities as an experimenter through rigorous aural attention. Hamilton contended that an accurate sense of hearing was essential to psychical research becoming recognized as truly scientific. Correlating what was heard in the seance with other psychical investigations revealed "similar fundamental characteristics" across the spectrum of paranormal phenomena that confirmed the sound's legitimacy as an empirical entity that could be appropriately quantified.[121] To make such connections tenable, Hamilton listened for the characteristics of the sound or voice, as well as identifying where it occurred in the room in proximity to the medium. Documenting the development of the voice, both automatic and direct, Hamilton noted its shifts in tenor, volume, and quality. The voice was initially "very indistinct" and "throaty" but became increasingly clear and skilful.[122] Similarly, he documented specific sorts of sounds that occurred throughout the sittings. Analyzing a "rustling" auditory effect similar to "the crumpling of silk or the creaking of new leather," which he apparently detected "from some point near the ankle of the medium," Hamilton compared it to the findings of Albert von Schrenck-Notzing, William Crookes, and William Crawford. Correlating such auditory stimuli to the mechanized inner workings of preternatural phenomena, he argued that standard features made themselves apparent in his own and other credible investigators' experiments. Testifying to "the same leading characteristics" in all of these hums, thuds, and reverberations, Hamilton asserted that such tests resulted in consistent phenomena that could be effectively appraised by scientific methods.[123]

The variety of noises, knocks, and creaks of the seance room shifted to a series of standard sounds. In the Boston seances, the first evidence of a supernatural presence came about through a telekinetic force applied to the table – a phenomenon that rather quickly developed into a systemized series of raps for the explicit purpose of communication.[124] As opposed to some incoherent knocks, such "table taps" proved instrumental to identifying the existence of various unseen intelligences at work in the seance room. "Real individuality shows in the taps," O'Hara Pincock related, "some quick, others heavily tap and some very heavy and strong."[125] Hamilton concurred, arguing along the lines of French physiologist and psychical researcher Charles Richet that the "reality of raps is of primary importance; and this phenomenon carries the implication of the whole of metaphysics."[126] Hamilton urged, again using the words of Richet, that "if it is established that mechanical vibrations can be produced in matter, and without contact, and that these vibrations are intelligent, we have truly a far-reaching fact that there are in the universe human or non-human intelligences that can act directly on matter."[127]

Although table taps were once a mysterious form of spirit communication, twentieth-century psychical researchers transformed them into indicators of demonstrable, repeatable, and mechanized paranormal effects. In addition to noting a number of mysterious "rustling," creaking, and crackling noises fairly early on in his investigations, Hamilton recorded the first instances of table rapping during his Winnipeg experiments in 1921. These spontaneous raps varied from "ticks which seemed to come from within the wood of the table, to loud knocks audible to all." This rapidly developed into a series of auditory codes, "vibrations that are not due to chance, but mean something." Hamilton reported how initial communications with discarnate intelligences progressed over time as the seemingly sporadic table taps were transformed into something coherent and understandable by aligning the citation of the alphabet with the taps so that they indicated a specific letter or phrase. Such audible messages, Hamilton contended, proved that these expressive energies "were not only intelligent, but in their choice of subject matter, of words, aptness of phrasing, correctness of spelling, they disclosed an intelligence considerable greater than shown by the medium in her normal condition."[128]

Effective auditory perception met with impediments in the seance room, including the limitations of the psychical researcher's body. Hearing came to be conceived of as a higher, disembodied form of perception, a quality only emphasized by psychical researchers' efforts to mechanize and standardize sounds. Listening to the various voices, taps, and rustles of the seance room proved impossible, however, if the examiner could not hear. This was by no means an unknown problem in the experiments. During experiments of January 1926, Crandon raised concerns about two John Hopkins University academics who were acting as the chosen personnel of an inquiry sponsored by the ASPR. One investigator was physics professor Robert W. Wood, and the other was psychology professor Knight Dunlap. Upon their initial and informal sitting, Crandon reported his immediate suspicion that "Wood could only hear in part the whispered voice of Walter, and Dunlap heard nothing." Both men confirmed the truth of the allegation and admitted that they suffered from hearing loss, "Wood saying he could hear the lightest whisper better than the strident whisper, and Dunlap saying that he had not heard a whisper since childhood."[129] For Crandon, this was the first of several objections to the experimenters, as they could not reliably detect one of the most significant aspects of the mediumship: Walter's voice. Crandon therefore "insisted upon the withdrawal of both Wood and Dunlap" from any commission of investigation of Margery's mediumship.[130]

The sense of hearing was not the only so-called heightened perception that collided with the limits of the body. Collecting visual evidence preoccupied many investigators and played a significant role throughout their experiments. Believing sight to be the highest sense most capable of relaying objective information from the surrounding environment, psychical researchers tried various methods to make the power of the spirits appear before their eyes and thus emulated a much larger trend in the natural sciences and modern society more broadly.[131] Sight and scientific rationalism were closely linked in the nineteenth and twentieth centuries, as scientific institutions heralded observation as one of the most preeminent necessities of its enterprise.[132] Vision attained what Mark Smith has called "a hegemonic position in Western culture," one that was set apart from and superior to the other senses.[133] As a consequence, "scientific visualism" in the twentieth century became an assumed marker of rational perspective.[134] Making things visible and observable thus played an intrinsic, albeit not unproblematic, role in structuring the science of the unseen.

Vision, much like hearing, met with a number of challenges in the seance room. In fact, since the nineteenth century, the seance had seemed plagued by conditions inimical to sight, from darkened rooms and enclosed cabinets to a number of misleading distractions, visual or otherwise.[135] This trend continued into the twentieth century. "Observation is fundamental to science," Henry Clay McComas exclaimed, a conviction that led him to lament how difficult it was to use sight for the examination of the paranormal: "They are not many situations so difficult for good observation as the séance." The chief of these was the "disturbing darkness," which, in the process of obstructing one's ability to see, aroused "haunting fears of the dark." Describing rooms "so completely dark ... that you could pass your hand immediately in front of your eyes and see no change in the darkness," McComas emphasized that this was a feature of the seance that the skilful psychical researcher needed to overcome.[136]

Eric J. Dingwall and William McDougall similarly found the blinding conditions of psychical investigation regrettable. As the few moments of clear-sightedness in their investigation of Margery were "brief," they were compelled to supplement scant visual evidence with both touch and hearing.[137] Their vision impaired, Dingwall and McDougall quite literally found themselves fumbling in the dark, a situation that they viewed as "a grave weakness."[138] If the happenings of the seance room could not be systematically observed, its phenomena could not be deemed legitimate by empirical standards. The ability to see with one's eyes those objects that were apart from the body supposedly meant that subjective compulsions could not taint the perception.[139] Yet this allegedly disembodied perception could not escape the carnality of the investigator, whose unaided faculty of sight proved wanting. As the following chapter details, investigators began to take steps to make vision truly disembodied through the employment of technology to supplement or supplant their limited visual perceptions.

***

The seance aroused a rich and varied array of perceptual encounters, from wafting perfumes to ringing electric bells to wet and slippery touches. Ghosts associated with the East, meanwhile, provided rich odours and an array of voices indicating their racial divergence. Encompassing and filling the mediums, the spirits transformed these uncanny instruments into incarnations of a sensuous and fully accessible other. The array of perceptions contributed to the belief in the phenomenon's materiality and, hence, its empirical value. Psychical investigators relied on such physicality to produce observable and measurable data that they deemed necessary for their pursuits to be considered scientific. Yet these sights, sounds, touches, smells, and tastes needed to be controlled and organized in order to reflect the standards of objectivity to which researchers fervently clung. Applying methods of aggressive, even violent, tactual control, interwar psychical researchers readied themselves to detect any suspicious odours and employed their own perceptions to counter the spurious sensations that the spirits eagerly produced. Despite the usefulness of touch and smell, investigators prized the so-called higher perceptions, which they deemed suited to collecting reliable information from the world around them. Hearing and vision became the ultimate means for spirits to be discerned, each allowing for observations to be made of phenomena remote from the body, thus ensuring the researcher's utter objectivity.[140] As Haraway explains, such was the defining quality of the "modest witness ... the legitimate and authorized ventriloquist for the object world, adding nothing from his mere opinions, from his biasing embodiment."[141] What the white male investigator perceived constituted what could be counted as reliable evidence, in contrast to supposedly superstitious, feminized sensory modes.

However, the embodiment of investigators would not so easily recede. When applying the perceptions conceived of as transcending the subjective body, psychical researchers found that the limits of their own corporeality betrayed them. Occasionally deaf and unable to see in pitch-black conditions, psychical researchers learned that the medium and the spirits were not the only ones plagued by excessive embodiment in the seance room. Such a predicament proved to be an extraordinary dilemma for investigators, who wished to transform the pursuit of the otherworld into an objective and calculable science comparable with other modern physical sciences. As a result, psychical researchers looked beyond their own limited and frail bodies to a series of technological apparatuses. Either enhancing or entirely replacing the imperfect perceptions of the investigators, such apparatuses held the promise of calculating the paranormal to make the objective character of such ghostly realities undeniable.

4
The Qualities of Quartz: Technology, Inscriptions, and Mechanizing Vision

Science in the narrow sense has not to do with immediate experiences, but with Nature, i.e. with empirical reality in space.
– Hans Driesch, "Walter Prince as a Scientist"[1]

To reach the standards of objectivity, psychical researchers fitted dark seance spaces with various instruments that would make psychical forces measurable, verifiable, and demonstrable. Whether common scientific devices or specially designed contraptions, these machines were intended to supplement, and sometimes replace altogether, the embodied and presumably feminine perceptions. The most important of these senses, as well as the one most in need of enhancement, was vision. Discontented with peering through the darkness with their unaided eyes, investigators employed several fixtures and apparatuses designed to see the mysteries of the seance room in a way that they could not. Hereward Carrington, Daniel Frost Comstock, Wallace K. Butler, Mark Richardson, T. Glen Hamilton, and others employed different forms of illumination, glass apparatuses, and in certain cases, an arsenal of cameras and lenses to make the invisible realms visible.

Increasing faith in and fixation upon machines and their demonstrable results fed into an expanding number of inscriptions, which in turn lent even greater credence to the idea that the metaphysical could be verified, measured, and objectified in the same manner as any other phenomenon of the natural world. Detailed records, alongside graphs, figures, and photographs, made up a textual and visual display of the methods psychical investigators used to verify and observe their unusual specimens. Employing machines and apparatuses that transformed the seance from an esoteric encounter into an empirical project, psychical scientists attempted to make the paranormal abide by scientific principles of measurement and observation.

Metaphysical Technology: Apparatuses of the Psychical Laboratory

Psychical investigators designed and built a vast array of apparatuses for the purpose of experimenting with paranormal phenomena. Convinced of the inadequacy of the unaided senses, they insisted that through such equipment metaphysical realities could be made tangible and thus analyzed according to the scientific method. These machines were not simply benign tools. They were, to use the insights of historian John Tresch, "bound up" with the conviction that nature, and even the capacity of humanity itself, could be reimagined through technological marvels.[2] Different devices played vital roles in the production of data in psychical laboratories, shaping the environment of the experiment, as well as the scientist himself. As psychologist and psychical researcher Henry Clay McComas asserted, the scientific man "spends his life with instruments of precision ... He works with them and they work upon him. Gradually he acquires the feeling that he must have accuracy and precision." No longer left to discern mysterious phenomena by his perceptions alone, the ideally masculine psychical researcher could now observe and classify "by means of machinery."[3]

Psychical researchers' confidence in machines reflected a wider trend in scientific ideology and society more broadly. Science and technology became critically linked in the interwar period. As a result, both were equally blamed for the mass death on the Western front in Britain and Europe more broadly.[4] North Americans, who were arguably more distant from the horrors of the trenches, proved to be much less ambivalent about technology. As scholar Glen Scott Allen argues in the context of the United States, "post–World War I technology was received with the same sort of unmitigated glee that had greeted the first Industrial Revolution."[5] The years between the world wars witnessed mass technological innovation in the home, in the city street, and in the laboratory. Hailing technology, even over and above scientific ideas, North Americans embraced mechanization as the means through which the wonders of the modern world could be fully realized.[6] Differing attitudes toward technology were evident, in part, among British and North American psychical researchers. Society of Psychical Research investigator Eric J. Dingwall, for instance, seemed less optimistic about the use of technology in the seance. Arguing that it could lead to a false sense of security, he warned others "not [to] be led away by elaborate apparatus."[7] In contrast, American-based McComas argued that technology guaranteed reliable results: "You can trust a machine for accuracy."[8]

Importantly, however, such disagreements did not typically express themselves in the form of polarizing debates over human versus technology. Lorraine Daston and Peter Galison have pointed out that confidence in machines to uncover the mysteries of the natural world was not infrequently coupled with a belief that the trained judgment of the professional scientist also played a role in the creation of knowledge.[9] Psychical researchers of the 1920s and 1930s similarly remained convinced that man and machine must work in sync so as to verify scientific facts about the natural and supernatural world. The presence of mechanical apparatuses and the proficiency of the investigator worked to defend psychical science from claims of irrational subjectivity, inadequate perceptions, or faulty instrumentation. Creating an ideal sensory experience that was at once distant from the body but skilfully employed, psychical scientists used devices to measure and observe the paranormal phenomena of the seance room.

Investigators used a number of common scientific instruments to make once-hidden phenomena decipherable, perceptible, and thus objective.[10] Several psychical scientists, for instance, applied the stethoscope to the bodies of mediums during certain experiments. In parallel with broader medical practices, these researchers employed the tool to reveal the hidden inner workings of the medium's body. Its auditory quality surmounted the invisibility of the spirits and the darkness of the seance room at once. Rather than solely examining the functioning of certain organs such as the heart and lungs, investigators sought to unveil the direct material effects of the paranormal. Monitoring the heart rate and respiration of the medium, the stethoscope exposed the physical nature of trance, shifting it from a mysterious event between medium and spirit to something that could be observed and measured through the investigator's skilful use of the instrument. The device may not have erased the embodiment of the experimenter, as its purpose was to enhance rather than replace auditory perception. However, psychical scientists argued that by working in sync with the mechanism, they could perceive the preternatural more precisely. In aiding the perception of paranormal phenomena, the device transformed the investigator and his relationship to the medium. Like the stethoscope's initial introduction to medical practice more widely, this instrument and others promised increased distance and thus objectivity in the relationship between investigator and medium.[11]

This principle still operated when psychical researchers used the stethoscope in both typical and atypical ways. To test whether the voice was generated from beyond the medium in cases of direct voice, some investigators argued that a stethoscope should be placed directly upon the medium's throat while the disembodied voice communicated from some other spot in the room. As opposed to measuring heart rate, the use of the device in these scenarios did not reveal a concern over the health of the medium but satisfied the investigator's desire to detect fraud or deceptive behaviour.[12] In each case, however, the psychical scientist created a similar distancing effect between himself and the medium, placing himself in an impersonal position of authority over a passive, typically feminine other. Whereas the scientist acted as the observer who scrutinized the medium's body, the medium, situated beneath the stethoscope, functioned as the specimen to be discerned and analyzed.

McComas attempted to detect the source of the voices generated by the medium William Cartheuser. In particular, McComas wanted to examine the voice of the spirit referred to as "Elsie." Since it seemed remarkable that this high-pitched and feminine voice could originate from the vocal chords of a man, McComas was intent on revealing its mysterious and unusual operations.[13] Applying the stethoscope directly to Cartheuser's neck, McComas listened to the medium's throat so that he "could hear his carotid artery and the sounds of his larynx as he breathed."[14] Due to moments when he could not detect these regular sounds issuing from Cartheuser's body, McComas assumed that the medium was pinching the tube. Finally, in great triumph, McComas recounted that he suddenly heard Cartheuser changing the register of his voice, leading McComas to claim that for this moment the medium had forgotten to pinch the tube as he had done previously.[15] McComas's use of the stethoscope, in addition to his keen scientific observation and powers of deduction, thus joined to observe and explain a phenomenon of a previously unknown origin.

Psychical researchers applied other instruments to make the paranormal perceptible and thus open to careful study, scrutiny, and meticulous measurement. A variety of weighing apparatuses became significant in several paranormal investigations. Typically using scales or balances to study telekinesis, psychical scientists proposed that once-mysterious happenings could be made tangible. Convinced that weighing devices were "capable of scientific definition, numerical evaluation and precise analysis and examination," investigators operated them to measure the force exerted by supernormal means.[16] Researchers insisted that mechanical measurements provided an empirical means of gathering objective information. In some instances, the device consisted of a set of balances in which weights would be placed on one side. The spirits would then be asked to balance the scales by psychic force.[17] In other situations, experiments involved a much larger apparatus capable of lifting amounts of thirty to forty pounds.[18]

In 1922, after a few months of testing the telekinetic powers of the medium Elizabeth Poole by manual means, Hamilton decided that he needed "weighing equipment" to "verify this phenomenon more fully." This apparatus consisted of a spring scale that would measure the force exerted, which one could then read on the indicator by means of a small red light illuminating the figures in the dark. A set of pulleys attached to the scale held four strong cords of rope that suspended the table to be tested (see Figure 5). To initiate the experiment, Poole touched the

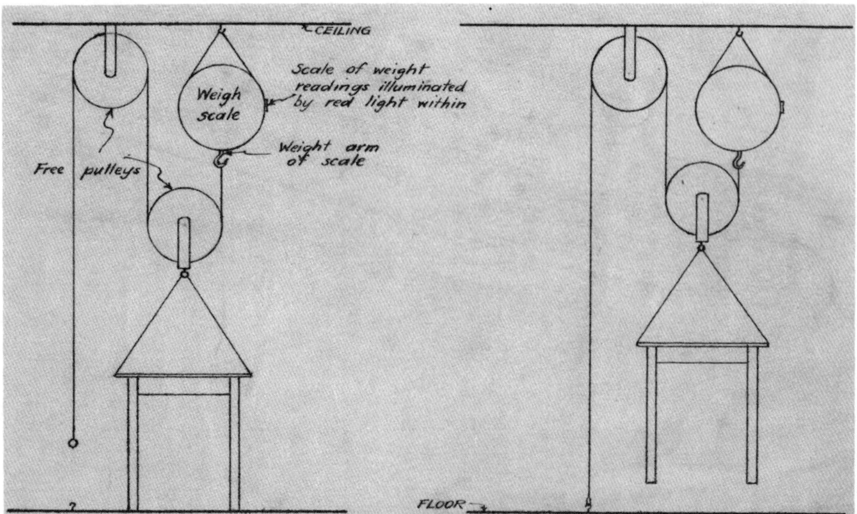

FIGURE 5 Diagram of weighing mechanism, c. 1922. A drawn rendering of the weighing equipment used to more accurately measure the paranormal force being applied during the experiments with Elizabeth Poole in the Hamiltons' psychical laboratory in Winnipeg, Manitoba.

tabletop for a brief period, much as she had in previous experiments. Once this was done, one of the investigators firmly grasped her hands above her head, while another asked the spirit to weigh down the table, taking careful note of the scale readings. Each time they conducted the experiment, Hamilton logged the weight measured. According to his records, the force increased from thirty-two to forty-six pounds in a little over a week.[19]

Through such experiments, the investigators at work in Hamilton's experimental room did much more than weigh mysterious forces. They transformed uncanny energies into something that could be discerned and measured, shifting paranormal power to the realm of the material. As was achieved with the stethoscope, hidden operations were rendered objective and accessible to scrutiny, comparable to any other phenomenon of the natural world. The promise that technology held for the transformation of psychical research into a legitimate science thus compelled investigators to employ additional apparatuses that were especially designed for the unique challenges they faced in the seance room.

Unlike the more recognizable devices of stethoscopes and scales, various mechanisms were specifically constructed for use within psychical laboratories. Investigators designed these apparatuses to maximize the potential of making otherworldly phenomena both verifiable and tangible in an attempt to surmount the unique difficulties they faced while quantifying the metaphysical. Mark Richardson, a co-investigator with Le Roi Crandon, designed one device to be used in the Boston seances with Margery, commonly referred to as the "voice-control machine." This apparatus came about as investigators tested and measured the direct voice of Walter. As with the stethoscope, experimenters in this context were convinced of the need to discover and validate the origin of the seemingly disembodied, "direct" voice. Tests of this order began in October 1923, when the medium filled her mouth with water at the precise moment when Walter's voice became audible. This test became increasingly precise, including measuring the amount of water Margery consumed and spat out throughout the duration of the experiment.[20] Other methods of ensuring that the voice was independent from the medium included manual control of the psychic's lips and tongue and insisting she place a toy balloon in her mouth.[21]

Despite these tests, Richardson remained convinced that a more mechanical means of testing Walter's independent voice needed to be developed. For this purpose, he designed an apparatus consisting of a large glass u-tube with a luminous float in each arm, connected to a corked bottle by a flexible hose. From this bottle, a total of six other hoses could be attached, each fitted with a glass mouthpiece at the opposite end. Richardson designed the mouthpiece with a flange and carefully placed holes so as to prevent those who held it in their mouth from inhaling or exhaling, or even adjusting their lips without moving the floats situated in the u-tube.[22]

After various attempts, the experimenters established a set of procedures to further ensure the validity of the experiment. Before the sitting, participants submitted to an inspection of their mouth and pharynx. Once everyone had taken their seat with mouthpieces in place, they turned off all lights. Each sitter then blew into the tube to lift the glowing float to a state of equilibrium. According to procedure, an appointed person briefly switched on the overhead red light to ensure that each sitter still had the mouthpiece in place. The hands of each person, particularly the medium, were held so as to prevent any tampering with the machine. Finally, investigators asked Walter to clearly enunciate an agreed-upon phrase that required a wide vocal range and opening of the mouth.[23] Sittings that involved just the medium and one investigator, or "one-man sittings," were especially sought after. With fewer people to account for, these types of sittings allegedly made the experiment "fool proof," at least according to J. Malcolm Bird.[24] In response to criticisms that a skilled operator could still speak while using the machine, Richardson redesigned the mouthpiece to ensure that it remained positioned at the centre of the mouth.[25]

Of all such devices intended to verify the spirits, perhaps one of the most popular in the psychical experiments under study was known as the "bell-box." Physicist and engineer Daniel Frost Comstock provided several apparatuses to be used in the Margery investigations, which he either purchased or designed and constructed himself.[26] He introduced the initial rough version of the bell-box to Boston's Lime Street seance room in 1924. A member of the investigative committee sponsored by *Scientific American,* Comstock aimed to scrutinize the phenomenon in a closed environment, which would limit the number of factors involved.[27]

For this purpose, Comstock came up with several models. One of the first consisted of a telegraph board, two dry battery cells, and an electric bell, which he wired so that the depression of either the key or the switch on the telegraph board would ring the bell. Once the spirit Walter demonstrated his ability to ring the bell in April 1924, they placed it within a sealed wooden box in anticipation that he would do so again but in much more controlled circumstances.[28] However, Bird deemed this apparatus inadequate, leading him to design and construct another model. He described the new version as consisting of "two eight-inch square boards" hinged together at one end, the other end being forced apart by a spring so that the boards formed a v-shape. Metal contacts were then fastened on the inside, serving as electrical conductors. Once appropriately wired, contact between the two boards against the force of the spring resulted in closing the circuit and ultimately ringing the bell. After some disagreement as to how it would be enclosed, the experimenters finally placed the bell and batteries within a locked wooden box. Running wire through small, drilled holes, they connected the batteries and bell to the contact boards, which were left outside and screwed in place at the top of the box. This new model proved so successful in leading to impressive telekinetic demonstrations that Bird referred to the bell-box as the "*pièce de resistance* of the Lime Street séance room." He argued that this apparatus and the now-measurable, standardized phenomena that surrounded it demonstrated the scientific and material significance of experiments with the paranormal: "that we are not dealing with miracles, but with orderly phenomena working under orderly laws."[29]

This version of the bell-box, referred to most commonly as the "*Scientific American* bell-box," became instrumental to verifying the disembodied personality of Walter. In fact, Walter and the bell-box remained so intimately tied that its very use acted as a trademark of the Walter personality well beyond the bounds of the Lime Street seance room. In 1928, four years after the introduction of the bell-box to the Margery investigations, T. Glen Hamilton built and incorporated a bell-box into his psychical experiments in Winnipeg that was "very similar in construction to the *Scientific American* bell-box employed in the Margery experiments."[30] He knew the design well, having been exposed to the device during his first experiences with the Margery mediumship in October 1925.[31] Not only did the "Fair Young Man" adeptly use the instrument during the Winnipeg experiments, but he also made demands as to how it should be operated and where it should be placed in proximity to the medium.[32] This led the Fair Young Man to reveal significant aspects of his personality, as well as details of his experiences with the Margery experiments in Boston. The spirit's apparent "character" and "mannerisms," in conjunction with his "methods of work and types of phenomena produced," closely aligned the Fair Young Man with the Walter personality of the Boston investigations and ultimately convinced Hamilton that he was dealing with the same lively, charismatic spirit. The use of the bell-box, therefore, provided Walter with an opportunity to demonstrate his presence across a remarkable amount of time and space.[33] Yet, as successful as this bell-box model proved to be in Boston and Winnipeg, it underwent a series of transformations in the years to come. Its redesign reflected psychical researchers' broader concerns over the dark conditions of the seance.

Light and Glass: Visualizing the Invisible

"It is a matter of universal experience that the greatest problem confronting the psychic investigator is to get the medium to sit in decent light," J. Malcolm Bird declared in his 1925 published account of the early experiments with Margery.[34] Attempts to make invisible phenomena visible met with a number of challenges in the psychical laboratory, much as the previous chapter identifies. The need to make the immaterial amenable to visual observation remained a consistent theme in numerous psychical investigations of the interwar period, directly shaping the use and design of various instruments.

Although unaided visual perception seemed nearly impossible, psychical scientists would not be dissuaded, supplementing and even replacing their own sight with a series of visualizing technologies. Their attempts were in keeping with the belief that machines could uncover the invisible forces at work in the natural world. As scholars John Tagg, Jonathan Crary, Suren Lalvani, and others have identified, a preoccupation with visualizing technologies gained incredible momentum from the nineteenth century onward.[35] The years after the fin-de-siècle were marked by an increase in the number and capability of devices designed to see what the human eye alone could not.[36] The first half of the twentieth century witnessed the development of the x-ray and high-speed photography, all of which exposed the once invisible to observation and control. Modern science and technology did not just chart, but also shaped, how time and space were understood and thought about, revealing the natural world to be much more complex than the naked eye could ever detect.[37]

Psychical scientists looked to technology to unveil invisible paranormal energies that reshaped how the contours of the natural world and life itself were conceived. Much as Hamilton argued, investigating paranormal phenomena exposed a world in which "energy forms the basis of life ... and so obliterates space as we know it in an ever-enfolding present obliterating time."[38] However, due to the nature of the phenomenon produced, complications did not end there. The mysterious forces under examination were said to be highly delicate. The intrusion of light into the seance room potentially damaged spirit productions, stunted the spirits' ability to operate, and even risked the well-being of the medium and sensitive sitters.[39] This especially seemed the case for condensations such as teleplasm or ectoplasm, which, according to proponents, were "soluble" in bright white light.[40] Certain investigators justified the fragility of the substance on the basis that many "physical and biological energies can only manifest themselves in darkness."[41] As a consequence, spiritualists and some psychical researchers asserted that the seance room needed to be kept in complete darkness, as it seemed to be "essential" to successful experimentation.[42]

Several investigators were not convinced by such opinions, however, and continued to insist that some form of illumination be introduced to the seance room. In a 1924 statement to Walter Franklin Prince, Comstock complained that he had "not yet seen ... such sufficiently definite and often repeated phenomena in the light." Here, Comstock repeated his general opposition to experiments carried out "in utter darkness," asserting instead that "every effort should be made to have the phenomena occur in a lighted room." Such conditions were the only way that "definite and often repeated" proof of paranormal phenomena could be obtained.[43]

In response to such conflicted interests, several psychical researchers attempted certain compromises. Notably, Comstock, as well as British psychical researcher Eric J. Dingwall, remained convinced that, despite the difficulties posed by white light for condensations and other phenomena, a genuine medium could be gradually trained to work in such conditions.[44] Accustoming the medium to bright white light should "be insisted upon at the very beginning," Dingwall argued, as continuous darkness during experiments was "wholly indefensible except in a few rare cases."[45] Others opted for the introduction of different degrees or forms of illumination that would prove less harmful to experimental conditions. Luminous bands attached to various apparatuses or the extremities of the medium sufficed for some. The aluminum seance trumpets of William Cartheuser commonly stood out in dark seance rooms due to bands of radium paint marking both ends.[46] Margery similarly agreed to wear luminous bands on her ankles, wrists, and head during certain experiments.[47]

Intermittent use of ultraviolet, infrared, or red light was another possible alternative. Tests revealed that each sort of light proved less destructive to paranormal phenomena than white, although no one could identify the precise reason why. Certain psychical scientists hypothesized that the wavelengths of each of these forms of illumination applied less strain upon teleplasmic condensations and paranormal phenomena, but little else seemed to be known.[48]

Operating on the premise that supernormal phenomena might actually be made visible through the use of ultraviolet light, individual investigators, such as McComas and Comstock, employed a mercury lamp to fill the seance room with these types of rays. Coupled with a camera equipped with a quartz lens, which was sensitive to a larger spectrum of light than ordinary lenses, they tested the possibility that supernormal forms could be detected.[49] Other investigators incorporated infrared light into their seance as a possible means of detecting phenomena. Hereward Carrington devised a method of testing Cartheuser by projecting an infrared beam of light between the medium and the trumpet. Carrington hoped to discover whether a solid substance could be detected acting on the trumpet. Arguably, such a body could be either something supernormal or merely the arm of the medium. By detecting its presence, Carrington could then further investigate the precise nature of the obstacle. As he passed the invisible infrared beam between and around the medium and trumpet, a small registering apparatus wired to the light revealed whether the beam had been obstructed.[50]

Whereas ultraviolet and infrared were used rather infrequently in psychical experiments, intermittent red light became a standard form of illumination in the seance. Several psychical researchers equipped both the main light of the experimental room and specific devices with a red bulb for periodic use during psychical experiments.[51] Overhead lamps and more portable flashlights were either covered with a red shade or fitted with red-tinted bulbs.[52] Comstock, Bird, and others attempted to incorporate various forms of red illumination during their investigations, including those that involved use of the bell-box. Initially, they placed a luminous plaque on the lever of the bell-box to reveal any attempt by a human hand or foot to ring the bell.[53] They then began to use red light for inspection of the bell-box both before and while it rang.[54] From August 1924 onward, the use of red light in the experiments with the bell-box became increasingly frequent, leading to various degrees of observable telekinetic phenomena.[55] As Bird's record for the seance of November 9, 1924 claimed, "Walter rang the bell, beginning and ending in red light ... The bell was immediately examined to show its freedom from external connections."[56]

Interested parties continued to express concern that despite the various options for at least certain kinds of light, experiments with the bell-box and other devices in the Margery seances often occurred during long stretches of complete darkness. Although not completely closed to the possibility that the spirits required such periods to produce phenomena, Dingwall warned that the insistence on darkness could very easily be viewed as "intervals not of incubation but of preparation."[57] Joseph B. Rhine was more damning, insisting that the only possible explanation for spans of imposed darkness was that it allowed for "fraudulent performances" and thus could be understood as little more than "a cover for trickery."[58]

To address such criticism, while also bypassing the damaging use of light in the seance room, others looked to the design and material of various apparatuses. Addressing the issue of inhibited observation, investigators incorporated apparatuses made of glass within the seance room, beginning with the seance cabinet. Since the late nineteenth century, spiritualists and psychical investigators had used cabinets during their seances. Usually a three-walled, wooden structure with a fabric curtain across the front, it served as a space into which the medium could retreat, allegedly enabling the spirits to produce various phenomena in sufficiently dark conditions.[59] As can be imagined, this situation allowed the medium to exercise a fair amount of freedom. Set apart and invisible from the sitters, she gained a brief reprieve from critical scrutiny.

For several psychical researchers, this situation was unacceptable. Hamilton, for one, limited the use of the front curtain by either pushing it aside or even removing it altogether during his Winnipeg experiments in an attempt to make the medium more consistently visible.[60] Mark Richardson went one step further during his collaborative experiments with Le Roi Crandon in Boston. In the summer months of 1925, Richardson first conceived of a new sort of transparent cabinet and introduced it to the Lime Street seance room later that year.[61] He constructed the frame, floor, and roof out of wood and nailed them together. For the sides, Richardson used sheets of glass, enabling any researcher to view the medium from various angles at all times. The cabinet came equipped with a hinged, glass door, which could be left open or shut and locked, enclosing the medium completely. The glass cabinet was six feet deep, three feet wide, and seven feet high.[62]

Although the space itself seemed to offer some latitude in movement, the controls with which the cabinet was fitted did not. Instead of relying upon "personal" or handheld control by another sitter, Richardson designed the glass cabinet to assert "a complete mechanical control of the Psychic's five extremities."[63] To achieve this effect, he carved three pairs of rectangular portals into the wooden framing on each side, allowing enough room for the medium's hands to extend out from wherever she situated herself, whether the back, centre, or front of the cabinet. Each of these portals came equipped with a hinged cover, which could be shut and sealed or left open to provide a shelf upon which the medium rested her hands. Eye bolts were firmly fastened above each portal on the exterior wall and floor of the cabinet, through which the investigators strung wires that were then used to tie the medium's arms and legs, thereby restricting her extremities and limiting general movement.[64] The glass cabinet isolated Margery from external tampering by enclosing both her and the phenomena that immediately surrounded her in a restrictive, even tortuous, manner, while still leaving her body exposed to the close scrutiny of the psychical scientists.

The introduction of the glass cabinet in the Margery experiments began to shape other devices used in Boston throughout the coming years, including the very popular bell-box. Despite the introduction of intermittent light, concerns continued to be raised that the bell-box needed to become more amenable to continuous observation. Dingwall raised objections to the *Scientific American* bell-box in December 1924, asserting that it lacked transparency. Its boxed design may have provided a more controlled space, but it did not facilitate clear observation. Dingwall therefore constructed alternative models that he believed to be more sufficient. So did a second investigative committee of Harvard academics in 1925. Each model attempted to address this issue of visibility either by narrowing the means of ringing the bell or by allowing Walter's methods to be made more observable.[65]

Expressions of concern regarding the bell-box's lack of transparency continued when electrical engineer Wallace K. Butler began attending the Margery seances. One of the first Boston sittings in which Butler participated involved the use of Richardson's transparent cabinet, and its design influenced a series of devices Butler soon invented for use within the Margery experiments. Known for his "inventive ability and genius for developing simple and effective apparatuses," Butler set himself to work on the problem of devising a series of contraptions that would demonstrate Walter's telekinetic abilities in a clearly observable, fraud-proof manner. On April 19, 1927, he brought four devices to the Lime Street seances: the "All-Glass Bell-Box," the "Eiffel Tower," the "Electric Tripod," and the "Selective Indicator."[66]

The all-glass bell-box in many regards seemed remarkably similar to the glass cabinet. In fact, other than scale, the chief difference between the two apparatuses was that in place of the medium, an electric bell sat at the centre of the box.[67] Insofar as female mediums were conceived of as complex technological apparatuses, such a switch might have seemed unexceptional to researchers.[68] Significantly, Butler's new bell-box addressed the issues of transparency and visibility that surrounded the original bell-box. Similar to the *Scientific American* bell-box, Butler's apparatus was fitted with an electric bell wired to two dry battery cells and wooden flappers that acted as a switch. In contrast to the enclosed, unobservable wooden encasement of that model, however, Butler's device was constructed completely of "sheet glass, cemented for rigidity within a light framework of angle brass." Its removable top – also glass – could be padlocked in place. Although open on one end, a series of glass partitions behind the initial glass wall with gaps on opposite ends created "a staggered or winding entrance to the actual interior of the box," making it virtually impossible for either a human hand or device to be inserted for the purposes of tampering with the mechanisms. A small red light inside the bell-box just above the wooden flaps illuminated the inner workings of the bell-box, particularly when experiments were conducted in the darkened conditions of the seance room.[69] Effectively, Butler had concocted an almost completely enclosed and secure device, which would be simultaneously visible to those beyond its glass frame. Much like the glass cabinet, Butler's new transparent bell-box allowed Walter to demonstrate his telekinetic abilities in a controlled space that was free of external infringement yet also amenable to scientific observation.

Following his innovative bell-box model, Butler designed a number of other devices that reflected the desire for telekinetic demonstration in an enclosed yet transparent and thus observable space. These inventions included the "The Little Theatre," which he introduced the following month. It consisted of a wooden frame measuring twenty inches long, fourteen inches high, and eight inches deep, propped up by a wood trestle. Butler used sheet glass to enclose the front, making the interior of the box clearly visible. Sheet metal covered the back, which Butler bent into "two vertical concavities" painted a reflective white. He installed upper and lower electric red lamps to ensure "an even lighting of the glass front," making the interior readily observable within the dark conditions of the seance room. A four-inch-wide plate glass shelf spanned the length of the little theatre along its lower half, supported by brass brackets. Various devices could be placed upon the shelf in order to subject their operations "to a maximum of observation and control." The little theatre became a main stage of sorts for the telekinetic manipulation of various devices. In addition to the Eiffel Tower and the electric tripod, other newer contraptions were placed within the little theatre, including yet another version of an electrical bell device, a set of glass balances, a slanted track referred to as the "Sisyphus," and an oversized compass dubbed the "Vane."[70] By working upon these various instruments, the spirit Walter demonstrated his telekinetic abilities, which necessitated dexterity and intelligence. The experiments involving the little theatre were significant, as fraudulent tampering would be extremely difficult, while providing verifiable and visible demonstrations of psychical powers.

An insistence upon transparent observation directly shaped the technology of the psychical laboratory. Psychical investigators introduced different forms of light and invented several transparent apparatuses to make visible what was once obscured. Yet one device not only enabled sight but also visualized what could not be seen by the human eye alone – a perfect mechanical medium for unveiling the invisible.

Spirit through the Lens: Cameras in the Seance

The camera was the most widely used technology of visualization in the seance room. Incorporated in paranormal experiments from Boston and New York to Winnipeg and St. Catharines, the photographic lens provided what Hamilton declared to be "the most valuable means at our disposal" of observing, analyzing, and objectifying the paranormal.[71] As scholar Suren Lalvani has identified, photography became a central way to enforce the "hegemony of vision" of the modern era. Beginning in the nineteenth century, governments and various institutions employed photographic technologies to construct "a visual ordering of the public domain," which in turn "signalled an intensification and heightened functioning of the instrumentalities of vision."[72] The camera's establishment as a prime visualizing technology faced numerous difficulties, but it still managed to receive "remarkable status as evidence and proof," even contributing to how evidence itself became defined.[73]

Social and anthropological sciences enthusiastically took up such devices as they expanded "their domain of expertise" over human bodies and the environment.[74] For the natural sciences, the camera gained a powerful empirical quality. Even that which was once invisible to the human eye, such as bacteria and other microscopic organisms, could now be seen. Moreover, the images of specimens could be disseminated to a wider public. As historian Jennifer Tucker has argued, these advancements did not mean that the evidential quality of particular photographs was not hotly contested. Those who disagreed over the merits of an image nevertheless hearkened to "the universality of photographic truth."[75]

Faith in the camera to reveal objective truth encouraged the boom in nineteenth-century spirit photography. To capture the "vital effluvia of the soul" on the photographic plate was to provide enduring evidence of life beyond the grave that could be viewed time and again by multiple witnesses.[76] Spirit photography notably emerged at the same time as new methods of photographic techniques, which enabled the manipulation of the image, including double printing and exposure. William Mumler, the first spirit photographer, launched his career in 1861 in Boston. His success revealed how financially rewarding the enterprise could be. When Mumler was arrested and charged with fraud in 1869, the dangers of unquestioned belief in the camera also became clear. Mumler was eventually acquitted, and numerous spirit photographers followed in his footsteps. The case, however, marked an important moment both for spirit photography and for photography more broadly. Although not the first time the evidential value of a photograph had been questioned, this event powerfully illustrated that the camera could, indeed, deceive.[77]

The idea that the camera could be manipulated to produce misleading images did not mean that faith in the device came to an end. Much as Miles Orvell, author of *The Real Thing: Imitation and Authenticity in American Culture,* has demonstrated, photography in the nineteenth century was primarily conceived of as an "artificial realism," the image providing "a representation of reality, a typification, a conscious simulacrum that elicited a willing suspension of disbelief."[78] The spirit photograph, although controversial, did not depart from that understanding enough to destabilize the authority of the camera. And in fact, by the end of the century, it had become popular for photographers to mimic the effect of a spirit photograph for the purpose of amusement.[79]

By the early twentieth century, the significance of the camera and its photographic results had shifted. No longer regarded as a means to produce a simulacrum of reality, the camera was conceived of as the "instrument of revelation" that formed how one viewed reality itself. But as Orvell also identifies, certain themes surrounding photography remained remarkably similar, namely "the persistent goal of representing some more intense, more authentic reality, beyond mere replication, something closer than 'realism' to 'the real thing' itself."[80] The advancement of photographic technology, from increasing portability to rapid shutter speeds, only furthered this conviction, leading famous twentieth-century photographer Edward Weston to grant the camera in the 1930s the ability to produce a "heightened sense of reality – a kind of super reality that reveals the vital essences of things."[81]

Twentieth-century psychical researchers took up the camera again, convinced of the technology's ability to present a more intense sense of reality by capturing energies that escaped the unaided human eye.[82] However, this time they were much more cautious, taking multiple views of the same phenomena and carefully guarding the development process against any possibility of deception. Since fraudulent spirit photographs continued to be made in the twentieth century, psychical scientists recognized the need to differentiate their experiments from dubious fakes.[83]

Comstock, who had already influenced the expanding mechanization of the seance room, introduced the camera to Crandon's Boston experiments. He commenced the first photographic phase of the Margery mediumship in May 1924, as yet one more initiative to make observation possible in the dark conditions of the seance room. Comstock hypothesized that paranormal condensations were invisible by reason that the light waves they emitted extended far beyond the spectrum of light visible to the human eye. Materializations might therefore become apparent through a combination of both ultraviolet light and quartz lenses, which, he and other investigators theorized, were more sensitive to light waves beyond the normal range of vision. Consequently, he obtained a "high-powered, quartz mercury-vapour lamp," which generated "a light peculiarly rich in ultra-violet wave lengths." He introduced this lamp, alongside two large cameras equipped with "high grade quartz lenses," to the Lime Street seances. With these elaborate apparatuses in place, the images he obtained did not lead to conclusive results, but they did exhibit a series of white, unidentifiable mists evident on the plates.[84]

When A.C. Lescarboura, managing editor of *Scientific American,* captured the same type of images without employing the mercury lamp, Comstock continued to photograph with the quartz lenses but used much less expensive red flashlights as illumination. On June 8, 1924, he positioned the two cameras side by side, aiming them at the medium who sat behind a set of balances. By this method, Comstock captured a "psychic cylinder" from two angles, revealing it to be three-dimensional. Perhaps just as important, his use of two cameras confirmed that the negatives themselves had not been tampered with. According to Bird, this event produced unequivocal evidence of supernormal forces at work within the seance room. Rendering the medium utterly passive as the "mere innocent catalyzing force," the camera and investigator had operated in sync to observe and record a paranormal substance of extraordinary character.[85] Supplanting the medium as the one who viewed the spirits, the camera enabled an objective rendering of paranormal forces to be perceived and analyzed by the expert eye of the psychical scientist.

Partly following Comstock's example, Eric J. Dingwall, William McDougall, R.W. Conant, Mark Richardson, and many others observed and recorded their various experiments with Margery by way of the camera. Each employed this visualizing device to capture different phenomena of either a telekinetic or teleplasmic character, documenting various points of the Margery case as they enfolded. Their efforts made unseen paranormal forces verifiable, demonstrable, and most significantly, visible through the lens of the camera. Despite this prolific photographic work, the Margery case was only one of many instances in which the camera played a central role in interwar psychical laboratories. At times, psychical researchers mentioned the camera only in passing, reflecting its assumed presence within the seance room and thus its place of prominence within psychical experiments.[86]

Hamilton became famous for the use of the camera in his experimental room. He believed that the camera, unlike the fallible human eye, provided the most reliable rendering of the otherworld. Not only was the mechanism untouched by human emotion, but it could also reveal what evaded the regular ocular senses. In a lecture delivered to the British Medical Association in 1930, Hamilton provided a general history of psychical research in which he emphatically expressed this conviction. Although he recognized previous attempts to practise scientific methods by several psychical researchers of the past, he identified the use of photography by German physician and psychical researcher Albert von Schrenck-Notzing between 1910 and 1912 as ground-breaking. Hamilton asserted that this practice marked the moment when "the reality of materializing substance was placed on a sound, factual basis."[87]

Hamilton did not use one or even two cameras but an arsenal of photographic apparatuses. He typically adjusted the selection of cameras in the seance room for each experiment. The exact number of cameras ranged from six to twelve on a given night, largely depending on the type of phenomena he expected to record.[88] He also frequently equipped the cameras with several sorts of lenses. Cameras of varying sizes and brands, with stereoscopic, rapid-rectilinear, portrait, wide-angle, anastigmat, or quartz lenses, made up his extensive collection, each intended for a particular purpose.[89] Whereas rectilinear and anastigmat lenses offered "fine sharpness of detail," wide-angle lenses made it possible to take "a general view of the group with experimental details not otherwise disclosed."[90] Meanwhile, stereoscopic lenses captured the "three-dimensional features" of certain condensations – an extremely valuable asset when attempting to verify the authenticity of materializations, as well as when analyzing and examining them. Much like other psychical scientists, Hamilton found quartz lenses extremely valuable to psychical experiments, as quartz seemed to "have a quality for recording things that are not registered by glass lenses." Despite lamenting the "softness of the negative," he recognized this special lens's unique ability to capture "ultra-violet rays," which led to "unusual disclosures" of paranormal phenomena.[91]

The various angles Hamilton was able to capture not only produced a detailed record but also differentiated his experiments from fraudulent spirit photographs of the nineteenth and twentieth centuries. Taking multiple photographs, each capturing experiments at different angles, Hamilton attempted to evade criticisms that his photographs were merely the product of darkroom fakery. He positioned each camera according to its specific function. Hamilton placed the wide-angle lens far back in the room, whereas he situated the cameras with rectilinear, quartz, or stereoscopic lenses "six to eight feet distant from the location of the object to be photographed" in order to produce "a maximum of size to the object upon the negative."[92] The numerous cameras "surround[ed] the point to be photographed in a fan shaped or semi-circular arrangement ... at various elevations from the floor."[93] He accomplished this arrangement with a set of wooden "double decked stands."[94]

Introducing the camera into the seance room was not without its challenges. The light required for photographs caused significant problems. So Hamilton came up with several strategies for limiting the degree and intensity of illumination to prevent undue harm to the psychical phenomena he wished to capture. To focus the cameras, Hamilton used a red lamp of "ruby glass sufficiently dense to be a safe light for ordinary photographic plates." Even so, he recommended that the lamp "be shaded entirely except for an opening on the side adjacent to the camera to be focused."[95] To create the required flash for the actual taking of the photograph, Hamilton used high-speed flash powder or electric flashlight bulbs that he permanently installed higher up on the wall of the room, above a closet door frame.[96] He expressed preference for these options over the floodlights typically employed for flash photography. The floodlights, he warned, were "too long in time of exposure for the substance to withstand."[97] Yet the alternatives also proved problematic, particularly flash powder, which Hamilton complained caused "an objectionable quantity of smoke." For this reason, Hamilton installed an electric fan and ventilator in the room that could drive the smoke out of the space. However, these additions limited the potential positioning of the apparatuses within the room, as the flash powder always needed to be close to the vent.[98]

Despite such difficulties and limitations, Hamilton remained convinced that the camera was essential to his experiments. Faith in not only the scientific method but also the camera itself epitomized a much wider belief in the technology's ability to represent a reality that was more radically real than what the human eye could ever perceive on its own. According to Hamilton, his psychical experiments in the seance room made visible "the reality of an unseen world of Life, Mind and Objective Activity ... captured by objective, unemotional cameras."[99]

Inscriptions: Photographs, Figures, and Documentation

Figures, diagrams, graphs, written records, and photographs accompanied the technology of the seance room, playing a fundamental role in the objectification of paranormal phenomena by verifying the experiments through a multilayered "visual display."[100] The constant generation of documentation provided psychical investigators with one of their most powerful tools for validating the data received through their apparatuses in the psychical laboratory and for disseminating their findings to the wider society. Bruno Latour describes the records amassed from scientific endeavours as "inscriptions." Technology, or "inscription devices," in turn provided the raw data that were then "cleaned, redrawn and displayed" on paper for wider circulation.[101] To render paranormal phenomena objective and tangible, psychical scientists compiled both textual and visual records, producing what Latour would identify as a "hybrid" between the world of paper and the world of the laboratory.[102] These inscriptions nevertheless remained unstable, resisting the labels imposed upon them while pointing to alternative meanings that undermined researchers' claims to power in the seance room.

A mass of records was generated by the Margery case. The amount and detail of the documentation only increased over time as investigations became more intense and controversial. During the early stages of the mediumship, several criticized Le Roi Crandon for his insufficient reporting, as he apparently omitted important details or, in certain cases, misled his readers. Consequently, Comstock and Bird reformed the documentary practices of the Margery experiments from 1924 onward, taking it upon themselves to compile detailed descriptions of the seances, to be accompanied by lists of measurements, figures, and photographs. Employing either a stenographer or a dictaphone, they meticulously documented the experimental procedures of each sitting in which they took part. The dictaphone was especially recognized for its value as a mechanical and thus objective recording device. The records it produced, "made audibly in the presence of all" the sitters at each investigation, were "regarded as official and binding."[103]

Compared to the investigators and experimenters of psychical activity examined in this study, T. Glen Hamilton produced an extraordinary number of inscriptions. From the early 1920s until his death in 1935, he created and collected voluminous reports of each paranormal experiment he conducted. The content of the records largely reflected his interests. As a medical doctor, he was fascinated with trance states, which he carefully analyzed and documented throughout his career as a psychical researcher. In 1927 Hamilton wrote that he had witnessed a total of 591 trances of the medium Elizabeth Poole alone, which he described as either "simple" or "compound" depending on how many stages ensued.[104]

Through a series of complex graphs, detailed reports, and photographs, Hamilton charted the onset and development of trance as it occurred within the seance room, categorizing each state according to its unique development and resultant phenomena. Although his graphs did not "show the more minute developments within each class," Hamilton reported that they did "reveal clearly the general trend, the diversity of output and the speed with which the mediumship moved to its peak."[105] Such an assessment seems modest when one reviews the precision with which he charted each trance. Hamilton assigned various physical and mental phenomena a number, the numerical progression paralleling the growing complexity of the trance phenomena. For instance, the early clairvoyance period prior to trance, which included visions of "objects and still-life," was assigned the number one. The number six, meanwhile, indicated a deep trance that displayed both writing and speaking on the part of the medium while compelled by apparently supernormal means. He graphed such phenomena along a time sequence of five-minute intervals, the dips and climaxes indicating movement from the categories of "normal consciousness" to "trance sleep" (see Figure 6).

Much as scholar Mary Poovey has identified with the historical emergence of the modern fact, such instances of concise, numerically ordered

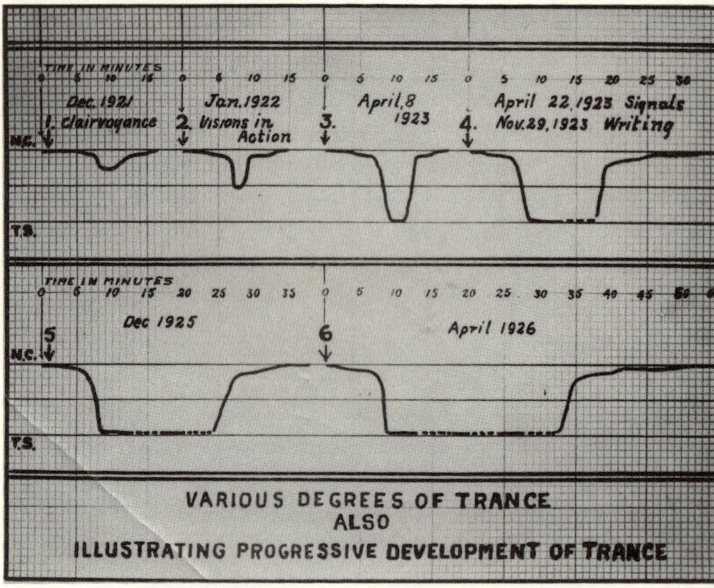

Figure 6 "Various Degrees of Trance," 1926. A diagram drawn by Dr. T.G. Hamilton that attempts to identify and track the various stages of trance experienced by the medium Elizabeth Poole during psychical experiments in Winnipeg, Manitoba, from December 1921 to April 1926.

classifications listed along a sequential graph signified "simple descriptors of phenomenal particulars" that "resist the biases that many people associate with conjecture or theory."[106] Separated from "excessive" narrative or long descriptions, the trance state appeared transfixed in these series of charts as an indisputable fact – an unbiased piece of data pointing to a clear, even predictable, conclusion.[107] The graphs provided an aesthetic of data through which the process of trance could be viewed and disseminated to a wider audience. Moreover, they provided visual evidence of paranormal phenomena that had been rationally organized using a series of lines, numbers, and categorical classifications. The objective nature of trance, now made apparent to the discerning eye, could not be disputed.

Since Elizabeth Poole seemed to have a gift for telekinesis, Hamilton also expressed a keen interest in this area for a time. Throughout these numerous experiments, Hamilton took careful notes, documenting not only alterations in the weight of the table and in the degree of force but also his assessments of the latter's various characteristics. Alongside these detailed measurements and descriptions, Hamilton also included photographs and diagrams of the phenomenon to reflect the arrangement of the personnel and apparatuses in the room, as well as the phenomenon

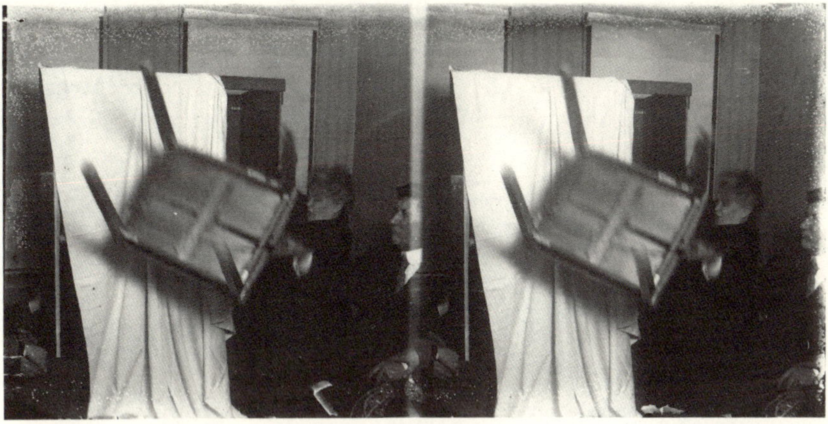

Figure 7 "Telekinesis: Levitation and Inversion," 1926. A photograph within an annotated photo album of the Hamiltons, featuring a levitating table experiment with the medium Elizabeth Poole in Winnipeg, Manitoba, in which the table is thrown violently out of the seance cabinet and into the air by an alleged paranormal force.

itself.[108] The photograph seemed especially suited to capturing the scientific rigour of the experiments. Mediums, flanked by cameras, were transformed at the moment of the flash into a mute object of study, or even at times into merely the unrecognized background of the phenomenon depicted. One photograph of Hamilton's, entitled "Telekinesis: Levitation and Inversion," portrays the table in motion during an experiment with telekinetics, a white sheet across the front of the cabinet accentuating the object's lines as it flies through the air. The medium Poole, meanwhile, fades into the dark shades of the wall toward the side of the photograph's frame, her name or presence left unmarked and unrecognized despite her central importance to the experiment itself (see Figure 7).

In other cases, Poole sat at the centre of the photograph, her presence and role made more definite within the image itself. This was the case with one 1926 photograph, in which she is prominently placed at the centre of the frame, her hand extending into the cabinet to rest upon the table, while her investigators look on. Yet even in this genre of photograph, Poole was still viewed as secondary, at least to her investigators. Included in an unpublished report entitled "Levitations by Psychic Force (Telekinesis)," the purpose of the photograph was to demonstrate the levitation of the table by psychical means. Labelled "Experiment E," it was listed alongside other images intent on demonstrating the experimental act and the objective phenomenon obtained. In the same way, the narrative surrounding the photograph revolved around the table and the investigators, ascribing to Poole a merely incidental part in the event at hand. As the caption of the image read, "Preliminary technique to securing independent movement of an enclosed table; the table is becoming psychically activated by contact with the medium's hand; the experimenters are ready with flashlight equipment and screen"[109] (see Figure 8). Similar to

FIGURE 8 "Experiment E," 1926. A photograph found in T.G. Hamilton's annotated photo album, "Levitations by Psychic Force (Telekinesis)." Elizabeth Poole is centrally positioned with her hand reaching into the open cabinet to touch the top of the table in order to "charge" it with the appropriate energies before levitation. T.G. Hamilton sits to her right and grasps her hand, while his left hand holds the mechanical switch to operate the cameras.

what literary scholar Clive Scott suggests of photographic captions more broadly, the description attached to the image attempted to diminish the "incoherence" and "randomness" of the photograph. Providing an objective and straightforward rendering of the experiment, investigators reconstructed the photographic moment of uncanny power as a rationally ordered event.[110]

The excessive description surrounding this photograph attempted to contain the image, reasserting the authority of the psychical researcher in the paranormal investigations. Yet the image itself hints at something altogether different. Despite being cast to the margins by the narrative, Poole sits at the centre of the frame, the power streaming from her providing the action, while the experimenters stand or sit back as little more than passive observers of a power clearly beyond their control. Her long, light-coloured dress casts Poole in aesthetic opposition to her investigators, who fade into the background as she charges the table full of paranormal energy.

Hamilton's photographs were useful for the purpose of visualizing the scientism of paranormal experiments. The images, however, were not always contained by the texts that surrounded them.[111] The empirical authority granted to these mechanically produced images was not impenetrable.[112] Although at times affirming the delicate project of psychical research, photographs could also display its vulnerabilities. Hamilton and other investigators went to great lengths in their attempts to manage unpredictable paranormal power, carefully charting its patterns and capturing its physical effects on film in order to make once-inexplicable energies accessible to rational and repeated observation. Yet, because of its instabilities, the image itself did not readily comply.

Only a handful of photographs survive from the seances conducted by the medium William Cartheuser and investigated by Jenny O'Hara Pincock in St. Catharines, Ontario. The few taken during a sitting within the space of the seance room were left undated, and the text labelling the images offers only a minimal explanation. One caption reads, "Scenes during Séance." The "minimal and non-interfering" label presents the content of the photograph it describes with transparent meaning.[113] The streaks of white that fill the space of the photograph depict paranormal manifestations moving from the spiritual to the material realities of the seance room. Within the frame of the photograph, the ghostly presence shifted from the realm of the invisible to a visible and even objective existence (see Figure 9). In contrast to the Hamilton photographs, however,

FIGURE 9 "Scenes during Séance," c. 1930. A photograph taken of a seance in St. Catharines, Ontario, in the home of Jenny O'Hara Pincock featuring an unidentified man, woman, and child. The bright white mark across the photograph is presumably paranormal phenomena.

the sparse explanation surrounding the image allows for alternative readings. The unpredictable power flowing freely through the room touches each sitter. A lone man toward the far edge of the frame is the only exception. He, unlike the women and the one child, sits apart and looks away from the seance circle, his gaze fixed on something outside the frame. He might be distracted by another person in the room, or perhaps he sees the source of energy that rests beyond the view of the camera. One cannot be sure. Rather than capturing preternatural power in its entirety, the photograph proves unable to portray all that is happening in the seance room. The image offers more questions than answers, its meaning as indefinite as the power that charges through it.

These proliferating "inscriptions" that investigators collected – whether reports, charts, lists, figures, diagrams, or photographs – all culminated in an authoritative self-verification of psychical phenomena. This vast and multifaceted assemblage of texts and images, in "mobilizing its own internal referent," to use Latour's terms, consequently "carries within itself its own verification."[114] Such inscriptions provided a scientific, objective basis on which to make judgments about the phenomena they illustrated, thereby drawing up one loop of the "knot" of scientific knowledge surrounding the paranormal.[115] At the same time, visual and textual records could also point to the instabilities inherent in the project of psychical research, their meanings as fluid as the psychic energies they were believed to capture.

***

Psychical investigators used both technology and extensive records to measure, observe, and verify paranormal phenomena. Employing instruments such as stethoscopes and scales alongside other pieces of equipment more exclusive to psychical science, investigators built an array of machines. The voice-control machines and mechanical bell-boxes all promised to transcend the limits of the body in order to perceive, measure, and objectify the paranormal. The invention and design of apparatuses specifically intended to promote observation only intensified this process. Making use of different forms of illumination, despite the potentially damaging effects of light upon their objects of study, psychical researchers demonstrated a growing dedication to a visual ethos. They placed mediums and machines within glass apparatuses that both limited the chance of fraud and increased the scientist's ability to observe psychical forces in a detached and objective fashion. Through the use of the camera, investigators produced images that transfixed the once invisible and intangible so that it could be observed again and again, providing the means by which the paranormal could be studied and verified by even those outside the seance walls. Photographs, accompanied by detailed records, graphs, figures, and other data, expanded a field of objectivity that increasingly encompassed the seance room – transforming inexplicable forces into systematic, repeatable, and measurable energies.

In this manner, the room of such experiments indeed became haunted but perhaps not in the way that investigators expected. A proliferating number of apparatuses and inscriptions paralleled and intermingled within increasingly ordered seance spaces. Yet the very same objects in which psychical researchers placed their trust threatened to betray them. Despite being buried beneath excessive description, the instability of the images produced in the seance room pointed to the possibility that paranormal powers could not be so easily categorized and controlled. As the next chapter demonstrates, the capacity of technology to visualize uncanny events also had its limitations, requiring investigators to take one more step in displacing the sensing, subjective body by looking to the mind. Psychical scientists coupled technology and scientific rigour with psychological explanation and medical analysis as a means to order and regulate the medium. Classifying her mind and body as fundamentally fragmented, they identified the psychic as a pathological subject. This fact alone made the medium a potentially destabilizing force within the seance – one that psychical scientists needed to observe vigilantly in order to keep her disintegrating self in check. Through the language of medical and psychological expertise, psychical researchers represented both the medium and the paranormal phenomena that emerged from her as in need of their scientific and medical intervention. In the process, investigators bolstered their own authority, while legitimizing their pursuit as a truly objective and scientific project.

5
Fragments of a Spectral Self: Psychology, Medicine, and Aberrant Souls

The ultimate fact is that the Universe includes Matter, Energy and *Mind.*
– Simon DeBrath, "Editorial Notes"[1]

Psychical researchers regarded the seance as a mentally and emotionally challenging arena in which "uncertainty and expectancy become contagious."[2] The dark conditions, the unexpected sensations, and the propensity to revive deep-seated fears were all elements that made the space as much a realm of the psyche as it was of the psychic. Technology may have been able to thrust physical paranormal phenomena onto visible planes, but it could not fully discern the minds at work in the seance. Taking one more step to diminish attention to their fallible embodiment, interwar psychical researchers represented themselves as experts in the field of psychology – an area of study that was booming in the years following the First World War. Employing a range of old and new theories of association, dissociation, and personality, investigators' preoccupation with the psyche extended from sitters and mediums to the spirits themselves. They carefully categorized and diagnosed uncanny and irrational minds, demonstrating their mastery over the disintegrating personality of the medium, as well as the phenomena that emerged from her. Phantoms could now be discerned in relation to the pathology of the medium.

The conviction that the mind of the medium was on the verge of fragmentation and mental illness directly affected how investigators observed her body. The mind and body of the medium could not be entirely separated, after all, for although psychical researchers aspired to an objective rationality untainted by the body, mediums could not escape the bounds of their own subjective materiality. Following this logic, psychical researchers demonstrated that the body of the medium was just as broken as her mind through a series of medical classifications that aligned the psychic with illness and disease. Psychological and medical methods adopted by psychical researchers interwove to contribute to deformed visions of another kind.

Associating Ghosts: Psychology, Personality, and Discarnate Intelligences

"No matter the angle from which we view the facts of Psychic Research," T. Glen Hamilton argued, "we are constantly faced with the presence of personalities behind and directing the phenomenon."[3] Perhaps unsurprisingly, psychological analysis played no small role in psychical experiments. Paranormal investigations of the 1920s and 1930s regularly involved scrutinizing the emotional states of the sitters, the physical and mental condition of the medium, as well as the distinct personality traits of individual spirits. Through such analysis, psychical scientists of the interwar period maintained the historical links between paranormal exploration and psychology.

The intimate relationship between the two fields is well documented by many who have noted psychologists' fascination with the paranormal, which in turn deeply influenced psychical research as it developed in the nineteenth century and beyond.[4] A number of the founders of psychology and psychoanalysis experimented with trance, somnambular, and dissociated states, including William James and Carl Jung.[5] This longstanding connection was fully intact by the interwar period. Episcopalian minister Rev. Dr. Elwood Worcester was a noted advocate of Christian psychotherapy while leading the Episcopalian healing ministry known as the Emmanuel Movement from 1906 onward.[6] Upon his retirement in 1929, his interest in experimental psychology and medicine shifted seamlessly to psychical research. He maintained in 1935 that since "psychic research concerns itself primarily with the resources and manifestations of the human spirit," it comprised "a department of psychology."[7]

Interwar psychical researchers were by no means alone in their interests in the psyche. Psychology, psychoanalysis, psychotherapy, and theories of consciousness obtained significant cultural power following the First World War. Theories of mind grabbed hold of the popular imagination during this period to contribute to radical new notions of the self that celebrated "values of autonomy and self-realization."[8] A shift from an emphasis on "character" to "personality" in the first half of the twentieth century indicated not simply a change in terminology but also the conceptual emergence of "a self for an industrialized and urbanized age: expressive, adaptive and morally unencumbered."[9]

In such a context, one goal of paranormal investigators was to examine the nature and characteristics of individual personalities through rigorous psychological analysis and testing. The minds they studied ranged from that of the medium to those of the spirits that emerged in the seance. Psychical researcher Hereward Carrington used word association tests to judge the validity and nature of discarnate personalities. Closely affiliated with Carl Jung, such tests were initially engineered to investigate the timing and nature of unmediated associative reactions to a series of specific words.[10] Jung conceived of the assessments as a means of plumbing the depths of individual personality, arguing that the act of association unearthed repressed desires and memories deep within the human ego.[11] Contemporaries soon acclaimed word association tests as the foundation upon which personality types could be identified and analyzed.[12]

Carrington used the same sort of associative tests when working with otherworldly intelligences, which other psychical researchers generally viewed as ground-breaking. Simon DeBrath, editor of the British journal *Psychic Science,* reported in January 1934 that Carrington's "experiments are perhaps the first really detailed instrumental investigations of the content of the sub-conscious that have ever been made." As DeBrath related, the significance of such spirit personality tests relied upon the conviction that "scientific proof of survival rests mainly upon the validity of the 'messages' which purport to come to us from some spiritual world, through the organisms of certain peculiarly endowed individuals called 'mediums.'"[13] Since spiritualist theories argued that a person retains "after death his memories, general ideas and associations, emotional reactions, etc.," it should be possible, Carrington argued, to apply personality tests to someone who had died, much as one would a living person.[14] Operating on the assumption "that every individual differs from all others in the structure of his mind," which makes one's "reactions to single words ... characteristic of each individual," Carrington had a Jungian-inspired list of 100 test words that he spoke to his ghostly subjects one after the other. He instructed the cooperating spirit "to speak the first word that comes to mind as soon as the stimulus-word is uttered, without thinking about it." Carrington postulated that the spirit's reactions to each word should be theoretically distinct "from the medium in her normal state, if he is really an independent entity, as he claims to be." Conversely, if the reactions could not be differentiated between spirit and medium, such a test also proved whether the spirit was in fact "merely a fragment of the medium's sub-consciousness."[15]

To further the scientific credibility of such experiments, Carrington employed a galvanometer, a machine that was intended to measure electrical current but could be used like a polygraph to measure emotional responses: "The greater the emotion aroused, the more current passes through the body, and the greater the swing of the needle as shown by the number of divisions on the scale."[16] Much as the previous chapter illustrates, the use of such technology enhanced the credibility of psychical experiments by eliminating the possibility of conscious fraud, or so Carrington argued: "No matter how hard he may try, the subject cannot influence the results materially. If a word or question arouses an emotion within him, then the instrument shows it, the galvanometer needle swings over and the variation is recorded."[17] Providing impersonal data distant from the psychical scientist, such technology simultaneously drew attention to the body of the medium as the galvanometer unerringly traced a series of electrical charges that conveyed "the slightest emotional changes in the subject."[18]

Carrington carried out such experiments for several years on different mediums and their acclaimed spirit controls. Reflecting upon word association and personality tests, in conjunction with numerous other experiments examining paranormal phenomena, he insisted that to validate "the persistence of individual human consciousnesses," evidence needed to be continually collected and then "tested and checked, whenever possible, by modern instruments of precision."[19] Social theorist Nikolas Rose, in his genealogy of psychological knowledge, argues that the "power of psychology" rested "in its promise to provide inscription devices ... rendering the human soul into thought in the form of calculable traces." Psychology as an idea and a practice aimed to reimagine and transcribe the previously unknowable inward self as a set of "diagnostic categories, evaluations, assessments, and tests," which in turn "constructed the subjective in a form in which it could be represented in classifications, in figures and quotients."[20] Psychical researchers such as Carrington used psychological tests and assessments to make the discarnate minds of the spirits materially measurable and quantifiable, ultimately hoping to "prove that mind can exist and function independent of a physical brain."[21]

Such an obsession with the personalities of spirits flowed directly from the influence of classical scholar and philosopher Frederick W.H. Myers. One of the founders of the Society for Psychical Research in Britain in 1882, Myers, in his magnum opus *Human Personality and Its Survival of Bodily Death,* first postulated the existence of other states of consciousness, which he referred to as the "subliminal self."[22] These alternative streams of consciousness potentially regulated internal organs and helped to process external social encounters, including the interpretation of facial expressions and gestures. Moreover, multiple levels of consciousness might also operate within the human psyche, leading to inexplicable states and abilities that demonstrated a hitherto untapped potential of the mind.[23]

Ann Taves has argued that Myers's hypotheses of the capability of human personality and the subliminal self "created a theoretical space ... through which influences beyond the individual, should they exist, might be able to manifest themselves."[24] Certain interwar psychical researchers took the ideas of Myers even further, using them to redefine the limits of the human mind. Myers's philosophy informed Hamilton, who insisted that both the physical and mental phenomena he meticulously charted and recorded "manifested the working of directing intelligences of a supernormal order. These intelligences claimed to be personalities who had survived death."[25] Hamilton expanded the bounds of human consciousness to extend well beyond death. Thus his argument that paranormal phenomena originated from "the psyche of the human being" in no way signified a purely materialist perspective.[26] Through his studies of trance and automatic writings, Hamilton became convinced that complex personalities impinged on the much more "simple" minds of his mediums. Creating "a type of cross-correspondence between two brains (or minds)," such otherworldly intelligences directly affected the "sensory centre and the motor centre" of the medium.[27] Summing up his observances of the mediumistic trance, Hamilton reported that both he and his fellow experimenters witnessed "the action of personality and disturbances of sensation, perception and behaviour, in accord with the laws of personality."[28] Insisting upon the unmistakable appearance of the famous personalities Robert Louis Stevenson, David Livingstone, William T. Stead, and Camille Flammarion, among others, Hamilton argued that the various trance communications revealed entities of "so complex a nature that they could not have arisen from subconscious influences lodged within the mind of our uneducated medium."[29]

The psychical investigator's drive to understand personality and plumb the depths of its capacities alternatively led to a discussion of its aberrations. Such could only be expected, as psychological assessments of personality relied on assertions of pathology to derive a set of normals.[30] The study of such disturbed minds resonated with those psychical researchers who applied psychology's findings while examining psychics, who rested precariously between pathology and another world.

"Trance Syndrome": Dissociation and the Pathological Medium

Elwood Worcester echoed the claims of many investigators and commentators when he wrote in the mid-1930s, "Of all the human sciences, abnormal psychology is the discipline most akin to Psychic Research."[31] Although numerous paranormal investigators exhibited a profound interest in personality and its potential to exist beyond known realms, it was through the study of pathology and abnormal psychology that psychical research gained its claim to fame. Unlike the discussions of personality, which concentrated chiefly on discarnate intelligences, most of the analyses of mental and emotional pathology were directed toward the medium. Subtly or not, investigators presumed that despite the centrality of the psychic's mind and body in experiments, these unique specimens harboured ingrained tendencies for deviancy. Even as investigators recognized the medium's gifts, they classified her as a "complicated problem in abnormal psychology."[32]

The conviction that the medium was innately pathological influenced investigators as they studied mediumistic faculties believed to rest upon the operations of consciousness – something that preoccupied psychologists and psychical researchers alike.[33] Psychologist and psychical researcher Gardner Murphy described the study of different shades of consciousness as being of "vital importance" yet largely unknown and little explained even after years of study and debate. Of particular importance to psychical research was whether, as Murphy asked, "the mind [can] be split in two?"[34] What he was explicitly referring to was the state of "dissociation." French psychologist Pierre Janet is credited with coining the term in 1887 to refer to a splintering of consciousness from specific memories and actions, leading to highly differentiated and at times erratic behaviour and thought within a single individual. Janet's theories laid the groundwork for the psychological study of duality, or doubling, of consciousness.[35] Although initially conceived of as applying to those under hypnotic suggestion, the designation soon also referred to somnambular and trance subjects.[36]

By the twentieth century, dissociation had become intimately linked to mediumship, acting as a central point of reference for psychical researchers as they analyzed various psychic states. In fact, interwar investigators viewed this splintering of mind as a necessary condition for the discarnate personalities to emerge from the medium. Hamilton and other psychical researchers insisted that trance "is always associated with complete dissociation of the medium's personality."[37] Presenting the clairvoyant's trance as a fundamental "duality" in which two sets of consciousness operated within one individual, the theory of dissociation proved particularly useful to psychical scientists. Although it could and did lead to purely secular explanations, the theory of dissociation also allowed for more metaphysical conclusions. As J. Malcolm Bird explained, the apparent "extra identity" could be nothing more than "a subconscious masquerade, wholly fictitious." Nevertheless, an equal amount of evidence, he insisted, pointed to "an actual possession from without" in certain cases of mediumship.[38] By referencing the theories of dissociation and relating mediumship to the occurrence of doubled, or fractured, minds, psychical researchers thus still clung to less materialist explanations, even as they naturalized the act of spirit possession. Insisting upon the extraordinary character of mediumship, investigators argued that phenomena such as automatic writing, direct voice, and other behavioural aspects of trance could still maintain their place alongside the objects of more mainstream scientific pursuits.

Psychical researchers used the explanatory power of dissociation and other psychological theories of personality splits to explain supernormal communications. Researchers pointed to the distinctiveness of certain spirits, expressed through the tone of voice, mannerisms, gestures, and evidential material conveyed, which were completely alien to the medium in her "normal" state. Carefully observing the myriad forms of "gestures, facial mobility, pantomimes, motor control of the entire organism," psychical researchers such as Bird and Le Roi Crandon insisted that trance states revealed complete and unequivocal "dualities" of personality altogether distinct from that of the medium.[39]

Dissociation took on this diagnostic role in relation to mediumship in order to rationalize the uncanny state. Despite contributing to the wider acceptance of psychic ability as something that could be analyzed according to psychological and medical explanation, the diagnosis of dissociation did so only by linking mediumship to mental illness. Disassociation also permitted psychical researchers to uphold the significance of the mediumship and, by extension, their own role as scientific professionals, even if they themselves were not convinced of the clairvoyant's performance. If circumstances made it difficult to claim certain expressions of psychic ability as authentic, they could still be viewed as valuable since they demonstrated the operations of dissociation within the human psyche.

When examining William Cartheuser in the late 1930s, Carrington suspected that the voices emerging from Cartheuser while he was in trance displayed too much of the medium's conscious personality and desires. As Carrington noted, Cartheuser's "comments on certain sitters, the conditions involved, his avid desire for money, his native tendency to scandal and what may be termed 'bathroom stories' all are manifested in his alleged communications."[40] Despite his critical outlook, Carrington described the medium as "simple, naive, and sincere," adding that he seemed to be "a curious mixture of naiveté and cunning." Carrington observed how Cartheuser experienced an "interesting transformation" during these seances from his supposed normal self into a person who was "more positive, more assured, more mentally agile." Cartheuser's jokes, answers, and conversation during a seance appeared "to be more shrewd and spontaneous than he would be capable of in normal life." Carrington ultimately judged Cartheuser's mediumship as illegitimate.[41] Rather than dismiss Cartheuser as an irrelevant fraud, however, Carrington insisted that his mediumship still held importance to psychology as a demonstration of a splintered consciousness. Carrington theorized that Cartheuser's "usual personality has been, to some extent at least, overshadowed by some other personality, more alert than his own ... some slight dissociation takes place."[42] Thus, even though Cartheuser's mediumship proved wanting, Carrington saw value in Cartheuser's case due to what it potentially revealed about the condition of dissociation. The diagnosis superseded the spirits in its fundamental importance. "From one point of view it is immaterial *what* they are," Carrington argued in regard to Cartheuser's spirit controls, Alice and Dr. Anderson. Even if Cartheuser's spirits were little more than "subconscious fabrications," they nevertheless "constitute an invaluable connecting link, without whose presence the medium could not adequately function."[43]

As the previous example suggests, dissociation was not wholly benign as a diagnosis. Psychical researchers represented the medium through the language of dissociation, split minds, and illness, regardless of whether they thought the demonstration of psychical abilities was genuine. The very performance of mediating between this world and the next connoted a fracturing of mind that even mediums and their spirits feared could lead to mental disintegration. Some argued that trance, the quintessential state of dissociation, ultimately led to "obsession," which caused a complete loss of control of one's faculties through either possession or an unhealthy fixating influence.[44] The way that psychical researchers identified and spoke of trance only furthered the connections between dissociation, trance, and pathology.

Although Hamilton viewed mediumistic trance in a fairly positive light, his method of classification connoted its pathological potential. Referring to the "trance syndrome," Hamilton described how he and "other medical men" observed and recognized the trance state as a "genuinely abnormal condition, exhibiting a symptom complex unmistakably distinctive."[45] Similar to descriptions of disease, Hamilton's account of the trance depicted it as an "invasion" by which forces external to the body overtook one's normal functioning.[46] Trance required that the medium's personality be "gradually broken down under protest" and thus linked this supernormal condition to a psychological breakdown.[47] He carried this association further by describing "four distinct periods" of trance as one would a physiological and mental illness: "the stage of onset; the stage of deep sleep; the stage of recovery; and at times, a post-trance subnormal stage."[48] Although eager to study its various phases, he recognized that some stages of trance paralleled hysteria and that only a medium who retained "integrity of personality" could safely enter and emerge from the dissociated state.[49]

For Hamilton and his wife, Lillian, the key to understanding trance was Myers's conception of human personality.[50] Although Myers's theories remained open to the possibilities of the paranormal, they also supported the idea that trance was an erosion of the psyche. As religious scholar Jeffrey J. Kripal argues, Myers entertained "a rhyming connection" between "abnormal psychology and supranormal psychology." Although dissociative consciousness allowed the human mind to access other realms, it also permitted the personality to fragment, leading to "abnormal, pathological or devolutive states." These two states could in fact exist within the same individual, revealing that "pathological and mystical states are *not* mutually exclusive."[51]

When investigators scrutinized mediums – a consistently feminized subject of study – the tendency to pathologize proved even more difficult to resist. Several scholars have argued that womanhood and illness were not only aligned but also conflated. Just as doctors linked the female body and its sexual organs to disease, so too did they treat the female psyche as a fertile breeding ground for emotional instabilities.[52] As literary critic

Elaine Showalter has demonstrated, a "fundamental alliance between 'woman' and 'madness'" placed women "on the side of irrationality," making insanity "a female malady."[53]

The cultural link between femininity and madness directly affected how mediums were perceived, irrespective of the clairvoyant's gender. Psychical researchers portrayed female and male psychics as unstable minds on the brink of mental, physical, and even sexual pathology. Not inconsequentially, a significant aspect of paranormal investigations involved the assessment and monitoring of the medium's mental situation. Eric J. Dingwall held a fairly positive attitude toward those mediums he regarded as genuine, yet he still insisted that they were essentially abnormal, with a "childlike nature" and a tendency for "waywardness." The astute investigator, he argued, must be aware of such emotional instabilities, which could inform the investigator how best to manage the medium during an experiment. Even if certain behaviour or a particular message from the medium seemed nonsensical, such occurrences could still reveal aspects of the medium's "mental processes of which you are not aware."[54] Dingwall's assessment of Margery reflected his convictions. His report of his investigation of her in Britain and then in the United States conveyed multiple details of Margery's background and emotional state, including whenever he noticed Margery seeming "ill and depressed," as well as whenever she complained of an unusual pain or anxiety.[55]

Dingwall was far from unique in his perspective, as many others who examined Margery made note of her precise state of mind at any given moment, even beyond the seance walls. Bird insisted that Margery remained composed even in difficult circumstances. In contrast, specific members of the second 1925 Harvard investigation reported that Margery suffered from "a very unstable mental situation" and thus required the direction of male expertise to restabilize her apparently fragile disposition.[56] Hamilton similarly insisted that closely monitoring the emotional and physical state of Margery and other mediums during each stage of trance was an intrinsic part of responsible investigation. Linking trance to hallucinations and other pathological mental conditions, he insisted that mediums could suffer from "nervous shock" if not observed, examined, and ordered by professional medical experts such as himself.[57]

Science writer Waldemar Kaempffert predicted in 1920 a future in which "the student of abnormal psychology and the scientific investigator of psychical phenomena will join hands."[58] His prediction in many ways came true. Correlating mediumship, pathology, and women, psychical researchers forged their method of investigation.

Classifications of dissociations, or "doubles," to diagnose the fracturing of identity and personality represented the minds of women and mediums as interchangeable, whether statistical or medical evidence supported such an association or not.[59] Consequently, certain psychical researchers embraced such diagnoses, erasing many of the potential tensions between their own field and psychological explanation. The link between the two fields, however, carried its own dangers, as the paranormal potentially became nothing more than an unbridled fragment of an incoherent and feminine personality.

Psychic Bodies and Spiritual Parts: Materializing Fragmentation of the Feminine

Using the language of abnormal psychology, psychical researchers gave order to the medium's disintegrating mind. Yet the medium's mind could not be viewed in isolation from her physicality. Whereas the psychical researcher constructed himself as objective and able to transcend the body, the medium could not escape her subjective embodiment. Rigorous examination of the psychic's mind thus resulted in a close monitoring of the psychic's body.

Michel Foucault and Nikolas Rose have argued that the body in the context of the modern era became "the space of origin and of distribution of disease" and thus represented "the field upon which psychological pathologies were to be observed."[60] Studiously examining various limbs, organs, and functions, investigators' physical examinations mirrored their psychological analyses, with both the mind and the body of largely female mediums being portrayed as divisible into parts that were disconnected and fallible rather than united within a whole and intact person. Psychical researchers were hardly alone in this tendency, as breaking the body into categorized pieces was one of the chief methods of asserting a scientific view of the body used by both scientists and medical practitioners. As James Opp asserts in articulating the conflicts between modern medicine and faith healing, "Medicine saw the body through its discrete parts" rather than through a whole person, regarding it as "atomized ... and divorced from any sense of personal subjectivity."[61] T. Glen Hamilton, Mark Richardson, and Le Roi Crandon, all very active medical doctors, keenly observed and recorded the separate parts and functions of the medium's body. From skin texture to breathing patterns, from the rate of the pulse and heart to isolated physical characteristics, the body of the psychic became as split and fragmented as the mind under the examining eye of the paranormal investigator.

Psychical researchers' obsession with the medium's physicality reflected a central principle of their science: paranormal phenomena remained dependent upon, and thus intimately entwined with, the biology of the clairvoyants. Notably, this conviction entailed the "ethical responsibility" of the investigator to watch over the medium and ensure her health and well-being.[62] Any inappropriate behaviour, excessive light, or rough handling potentially endangered both the mental and physical well-being of the psychic. As Jenny O'Hara Pincock warned, "the life of the human instrument may be in peril by the slightest mis-move on the part of the inexperienced researcher."[63] But proclamations of such an ethical imperative to safeguard the medium simultaneously affirmed that, much as the medium's mind seemed in constant danger of pathological illness, so too did her body. Both required close monitoring by trained and predominantly male professionals. Through this lens of sickness and medical expertise, psychical investigators in turn justified their access to and close examination of the medium.

Like other psychical researchers, Hamilton felt obligated to watch over mediums and ensure appropriate care throughout the experiments – a concern that conveniently overlapped with his devotion to empirical methods.[64] Both priorities required the production of lengthy medical reports. Hamilton observed mediums' physical characteristics, responses, respiration levels, and pulse and heart rates, assigning each to the different stages of trance that he identified. Every stage consisted of both "subjective" and "objective" elements – again revealing psychical researchers' conviction that the physical and mental worlds of the psychic were linked. The objective symptoms related by Hamilton comprised a broad range of reactions, including "excitement, psycho-motor retardation, catatonic rigidity, complete relaxation, the deep trance sleep and various automatisms."[65] He took careful note of breathing patterns, hand gestures, muscle tone, and any unusual or erratic movements.[66] Trance was physiological as much as it was psychological, involving symptoms ranging from "frequent spasmodic movements" to extreme muscular "relaxation" or "rigidity," from "disorientation" to "a condition of apparent stupor" and "cataleptic states."[67]

Hamilton insisted that such symptoms did not happen randomly. Rather, trance was a process with distinctive and recognizable stages – one that resembled a medical practice with which he was very familiar.[68] As he explained in his lengthy records pertaining to his mediums' trance states,

"Trance bears a close analogy to anaesthesia. The onset is very similar and its effects on the medium are much the same."[69] Anaesthesia became intimately associated with modern medicine and surgery – contributing to many medical and surgical successes, even as it limited patients' control over their own health.[70] As a medical doctor and surgeon, Hamilton knew anaesthesia well. Trance, it would seem, was a spiritual form of the practice. And like anaesthesia, it allowed for a radical penetration of external influence upon the sedate body of the psychic. Hamilton approached trance as a state by which predominantly masculine scientific expertise gained access to the inner workings of clairvoyant bodies and, in turn, to the spirit personalities that possessed them. Depending on the degree of trance, either "certain parts" or "the whole nervous system" reacted to the possessive powers of the mysterious intelligences, revealing an "overwhelming influence" that affected not just the mind but also the entire body of the psychic.[71]

Hamilton's reports, complete with brief staccato phrasing emblematic of the medical science to which he subscribed, reflected his profound fascination with and analyses of spirit-to-body relations. "Complete anaesthesia. Mouth and chin rigid. Abdominal pains soft. Respirations subdued. Hand is spastic. Pulse about 82. Knee reflexes pronounced," he reported after a 1931 examination of Susan Marshall, a Scottish, working-class medium who was referred to by the pseudonym "Mercedes" and was the sister-in-law of the medium Mary Marshall.[72] Her name and background were incidental to these reports, whereas her body remained all-consuming.

Susan Marshall's ability to take on a position of extreme passivity was viewed by Hamilton and others as extraordinary, even in comparison to other clairvoyants. More than Elizabeth Poole or Mary Marshall, Susan Marshall's deep state of trance was profound, especially when the spirits called for a materialization. When reaching a certain point in trance, different parts of Susan Marshall's body became rigid and insensitive. According to Hamilton, her symptoms resembled other conditions such as "severe toxic conditions," "cataleptic states," and even "rigour mortis." Hamilton fervently monitored and tested Susan Marshall while she remained unconscious, taking note of any extraordinary features and the depth of her trance.[73] Attempting to shift her jaws, bend her legs, and even prod her eyes, he systematically reported and analyzed every detail of her body's anaesthetic-like response. "Completely anaesthetic," he reported on October 5, 1930. "Jaws rigidly fixed. Eyeball is insensitive. No twingings of the eyelids. No sign of respiration. Pulse about 78.

Rigidity at pelvis and knees absolute. Arms now more limp. Chin still set ... Completely insensible."[74]

This same degree of surveillance continued as Susan Marshall regained normal consciousness. Apparently, her deep levels of trance caused her to feel the effects more markedly and for longer periods than did other mediums. As Susan Marshall slowly emerged from incoherency, her body still remained open for careful observations, tests, and prods. As Hamilton reported, "the skin felt cold to the touch; the face was pale and showed signs of bodily fatigue for some time after the trance proper was over."[75]

While the experiments progressed, however, the fragmenting of the medium's body into categorized pieces quite possibly contributed to phenomena of an entirely new kind. Beautiful and full-bodied apparitions had appeared in the seance room since the late nineteenth century. The ghostly bodies of the interwar period, in contrast, appeared incomplete and malformed. Various body parts and truncated torsos made their way into the 1920s and 1930s psychical laboratory, paralleling the ever-more-fractured body and mind of the psychic.

The apparition of two ghostly forms in Hamilton's Winnipeg investigations illustrated the pervading presence of the fragmented body in interwar seances. Both emerged in association with Susan Marshall, and they were, not inconsequently, gendered female. The first occurred on October 27, 1929. On this night, the spirit Walter claimed that he was trying "to build a material body ... It will not be a complete body – only part of a body." The form would result in "a good photograph of your good little friend Lucy" – one of Susan Marshall's spirit controls.[76] For several months after, Walter and the other spirits continued to refer to a coming materialization. Then on March 10, 1930, "The promised form appears. An imitative body of a woman seated on a chair," which fellow experimenter H.A. Reed described as "a white bust."[77]

Rather than a whole human form, this figure looked "moulded," with no hands or legs. It was a shrunken and misshapen white mass in the rough form of a head and torso, the only clear human resemblance being a small woman's face looking out from beneath cottony layers (see Figure 10). Hardly decipherable yet nevertheless praised as "the Great Lucy Materialization," it reflected and reiterated the fragmentation of the female body – a theme that persisted in these series of interwar seances.[78] Yet this was only the first of two bodily apparitions that rendered the spiritual envisioning of the feminine body as fractured as the earthly.

The second materialization revolved around yet another spirit control of Susan Marshall, referred to as "Katie King." Unlike "Lucy," who was

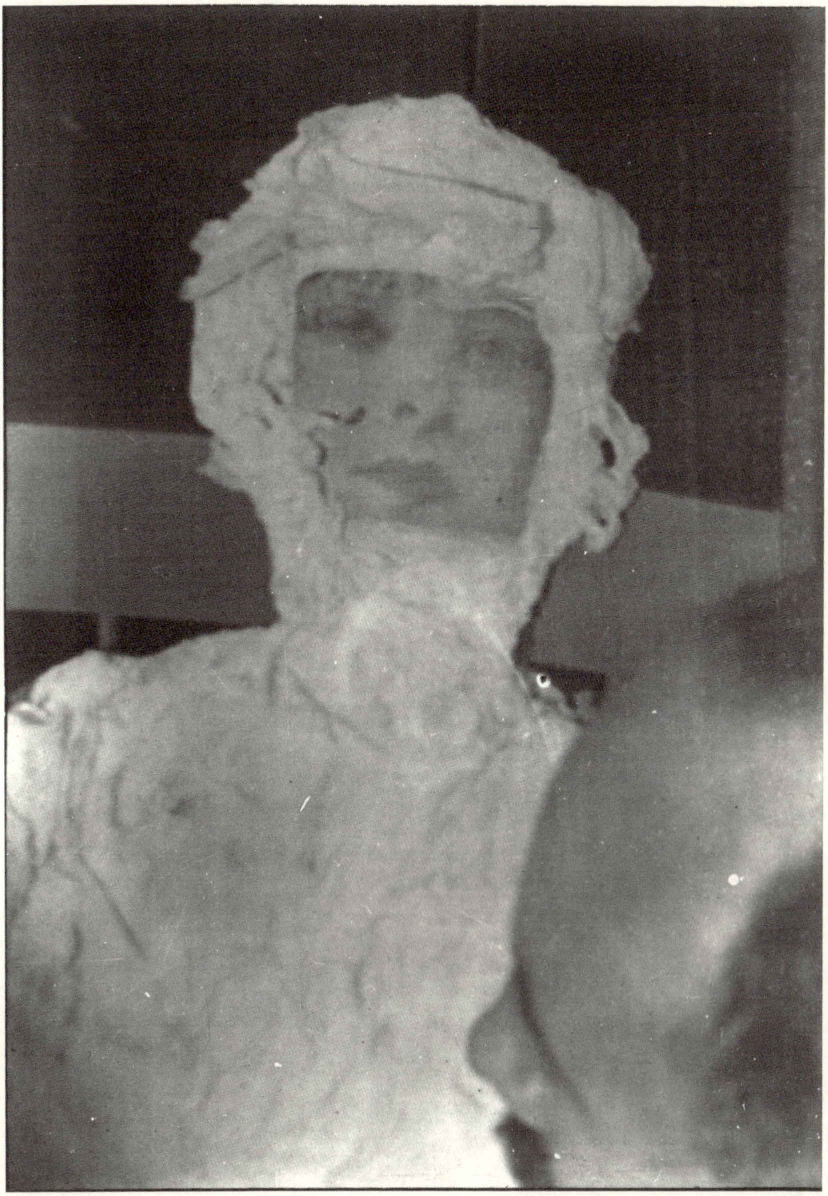

FIGURE 10 "Lucy, Said to Be Dead 100 Years," c. 1930. A photograph taken in the Hamiltons' psychical laboratory in Winnipeg, Manitoba, depicting a materialization of the spirit "Lucy" who spoke though the medium Susan Marshall. The face of the spirit stares directly at the camera, looking out through a white mass roughly moulded in the shape of a human figure.

relatively unknown, Katie King had a rather impressive lineage among those who sought the spirits. Once known as "Annie Owen Morgan," she was reportedly the spiritual daughter of John King, a notorious buccaneer who was known on earth as Sir Henry Owen Morgan. In nineteenth-century seances John King had been an exceptionally popular spirit among mediums, a figure whom Alex Owen claims "dominated nineteenth-century spiritualism."[79] The fame of his spirit daughter Katie heightened in the early 1870s when she appeared through the young British medium Florence Cook.[80] It was under the experimental guidance of renowned chemist William Crookes that the spirit of Katie King came into her own. During these lengthy experiments, Crookes performed various tests of authenticity, including photographing the apparition of Katie as it appeared to him, beautiful, full-bodied, and dressed in a flowing white gown.[81]

Sixty years later, Katie King resurfaced in the Hamilton experiments, or at least in parts. Her face, head, and hair, as well as veil and garment, made several appearances over the course of a few years. However, on February 21, 1932, Katie announced that she would make one more attempt to "appear in objective form."[82] Later, even the spirit of the now-deceased William Crookes came with promises to help them with the experiment.[83] Predictions ensued that the resulting apparition would be "brilliantly white and appear to be alive" – a description that seemed to correspond to Crookes's 1870 photographs.[84]

The actual appearance failed to live up to such standards. On April 23, 1933, an extraordinary phenomenon did occur. Hamilton reported, "Nothing like this ever before photographed in psychical research investigations." However, it did not resemble earlier appearances of Katie King, nor did it match the grand predictions that had preceded it. "The materialization appears. It is apparently lifeless."[85] As opposed to the graceful Katie King of the Crookes investigations, this apparition emerged from the spirit world as a floating, white, malformed torso – "a figure but without details."[86] In the photograph, a detailed but half-obscured face of a woman stares out from beneath layers of white, cottony material. A tattered veil enwraps the armless and roughly hewn form, falling down below the trunk over what looks to be shrunken legs cut off below the knees (see Figure 11). This climactic moment of Katie appearing in the Hamilton seances conveyed a stark revelation. Once more, femininity was made manifest in the seance – utterly broken and deformed.

Misshapen apparitions emulated the fragmented mind and body of the medium. Foucault has argued that dissection occurred not only at

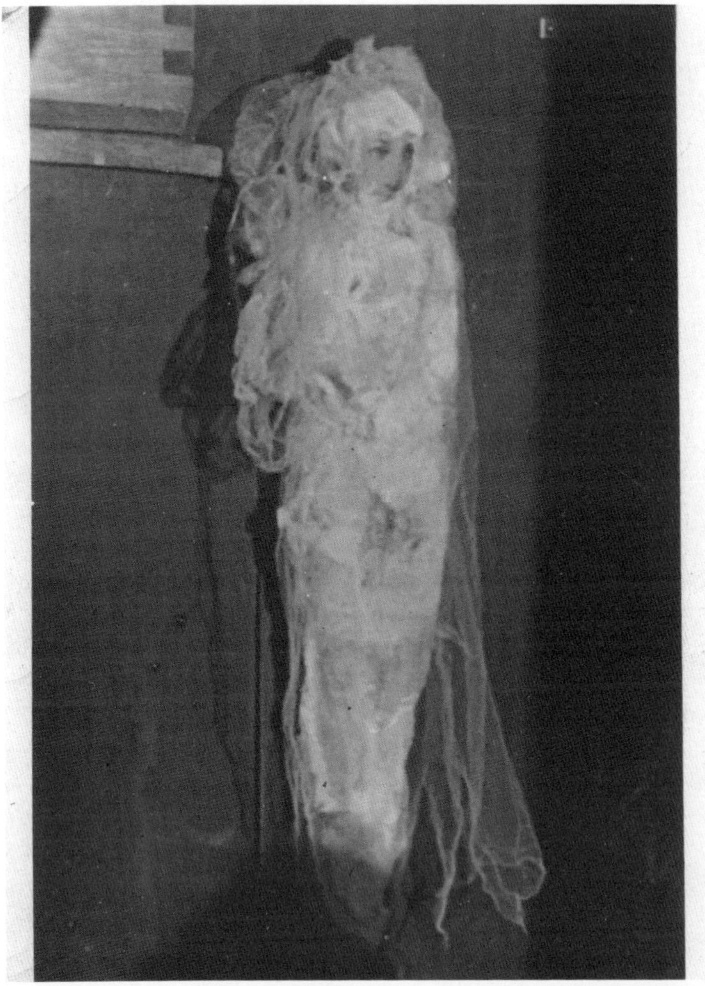

FIGURE 11 "The Shell," April 23, 1933. The much-anticipated appearance of the famous spirit "Katie King" in the Hamiltons' psychical laboratory in Winnipeg, Manitoba. Like the previous Lucy materialization, the face is only partially visable, looking out through a roughly hewn cottony mass that resembles a human figure cut off at the knees.

death but also increasingly in life through the discrete categorization of body parts and the symptoms that comprised the diseased body.[87] This physical and analytical act of breaking the body and its functions into isolated parts reconfigured death as "multiple and dispersed in time," as opposed to an "absolute, privileged point at which time stops." Dissection revealed the processes of the living, while removing death's "opaque character," which had once differentiated it from life.[88] The analytical dissection of the medium's body, both visually and supernaturally, aligned her with the broken ghostly bodies. Blurring the living with the dead, such strange visions of the seance reaffirmed the disintegration of the female body. The fragmentation of the medium defined her pathology, but it also denoted the domination of her physical and mental self by another. Psychical researchers legitimized their control over the psychic's body by segregating the clairvoyant's biological parts and responses into analytical and categorized pieces. The spirits that emerged reflected this systematic act of division, materializing in forms that were as fractured and as incoherent as the mediums from which they came.

***

Whether investigators were testing spirit personalities, assessing dissociated minds, or monitoring psychic bodies, it seemed quite evident that "psychological influences were at work."[89] Wielding the language, theories, and methods of experimental psychology, interwar psychical researchers studied the "abnormal" elements present in the seance room, which pointed to intrinsically feminine pathologies. Dissociated and fractured minds led to fragmented bodies and, in turn, to broken visions of an extraordinary kind. In the realm of the psychical laboratory, pathology, female bodies, and mental and physical fragmentation became intertwined concepts, which the appearance of ethereal, impartial bodies only made evident.

In part, investigating the paranormal opened up new vistas of possibility for the human mind. As researcher Simon DeBrath claimed, "there resides in the human mind, strange powers independent of time, space and normal channels of knowledge."[90] The application of Jungian-style word association and other psychological tests was aimed at uncovering this potential but always with reference to something abnormal that needed to be categorized according to psychological and medical expertise. Such rhetoric legitimized the psychical researcher's authority over not just the mental state of the medium but also her body. A mix of rigid examination, classification, and regulation of psychic minds and bodies revealed how mediums remained always possessed – either by the scientists who studied them or by the spirits who controlled them.

Mutilated and truncated spirits made the fragmentation of women perceptible, as they emerged from the same mediums who exhibited such a fracturing of the psyche. Indeed, these ghosts and the strange substances to which they gave rise provided an ethereal mirror of those who sat in the seance circle. As the mind of the medium materialized, so too did an image of the psychical scientist – rational, precise, and untainted by subjective embodiment. These reflections also became ghastly, however, revealing the instabilities of identity, gender, sex, and the body.

6

Teleplasmic Mechanics: Spirit Scientists and Vital Technologies

The question is nothing less than this: does the psychic in general – that is, the spirit, or the unconscious – arise in us, or is the psyche, in the early stages of consciousness, actually outside us in the form of arbitrary powers with intensions of their own?
– Carl Jung, *Modern Man in Search of a Soul*[1]

A mysterious substance known as teleplasm or ectoplasm engrossed several psychical researchers in the interwar era. Defined as a condensation of paranormal activity, teleplasm took on multiple textures and shapes depending on the experiment. These wet, misshapen, and all-too-biological psychic structures held the promise of affirming the science of psychical research, even as they skewed the boundaries between investigator and object. At once material, organic, and spiritual, teleplasm was simultaneously pliable matter and a supernormal agent, encapsulating Bruno Latour's insistence upon the "many metaphysical shades between full causality and sheer inexistence."[2] As the substance shifted from body to machine, teleplasm indicated that the investigators were not the only ones in the seance undertaking, and even transforming, the scientific project as it was conceived of in the interwar era. Ghostly scientists mirrored back an image of the ideal psychical investigator and gained extraordinary power while orchestrating experiments. Clairvoyants performed as delicate machines who needed to be properly handled if experiments were to succeed, only to destabilize gendered norms and radically upturn sexual hierarchies. Teleplasm, spirit, and medium thus combined to subvert the authority of psychical investigators and redefine what it meant to be scientific in the seance room.

Anatomy of Another Plane: The Vitality of Teleplasm

Psychical investigators went to great lengths to examine "an energy product not yet known to science," referred to as teleplasm or ectoplasm.[3] Since teleplasm was a hybrid between material substance and spiritual energy, investigators theorized that it was instrumental to accomplishing many of the feats of the seance room, in either invisible or visible form.[4] Teleplasm had already made its debut in nineteenth-century seance rooms, yet from the fin-de-siècle onward, psychical researchers focused upon the substance to realign their pursuits with contemporary and emergent scientific ideas.[5] This postulation continued in the following decades.

T. Glen Hamilton described the substance by citing recent scientific theories regarding energy and matter: "Matter must be regarded as stabilized energy and, as such, variable in its various combinations. Teleplasm is probably a very unstable combination of matter plus energy taken from the medium and sitters and is therefore quasi material."[6] Perhaps more than any other phenomenon in the interwar seance room, teleplasm was conceptualized by psychical researchers in a way that indicated their pursuits were not some throwback to an outmoded perception of science. Rather, the study of the substance's form and behaviour contributed to how energy, matter, and biological life were being reconceived in the first half of the twentieth century. Its significance, as a result, was not lost on Harvey Agnew, secretary of the Canadian Medical Association and fellow experimenter with Hamilton. Writing in the spring of 1931 to Hamilton, Agnew explained, "The replacement of the old laws of Newton and others as we understood them by the modern conception of wave mechanics made possible by the present knowledge of the construction of the atom, should be one of the strongest confirming arguments in favour of the acceptance of your work."[7]

As opposed to understanding teleplasm as an ethereal compound like their Victorian predecessors had done, certain interwar investigators recast the psychic structures they encountered in light of twentieth-century theories of atomic energy. Scientist and spirit concurred that teleplasm could take on multiple "forms and consistencies," either "vaporous or solid."[8] The spirit Walter outlined three general types, categorized according to the degree and kind of light they could withstand.

One remained "invisible" to human perception, being "entirely unable to bear any light." Another was invisible to the sitters but nevertheless "capable of being recorded on the photographic plate" as well as "able to endure red light." A third consisted of "sufficient solidity to be visible and palpable in red light and even in white flashlight."[9] Psychical researchers hypothesized that teleplasm took the different forms it did based on its internal atomic structure, which dictated how it responded to light. Agnew postulated that teleplasm's sensitivity to illumination could be accounted for by considering current physics research. Experiments of the 1920s and 1930s, he pointed out, indicated that the orbit of electrons widened around a proton when exposed to strong light, which created the possibility that electrons may escape, leading to the atom's disintegration. This finding could very well explain the behaviour of teleplasm: "As the atoms in the teleplasm are undergoing unusual changes and have not yet become stabilized, it would be very easy to see in this regard a possible explanation for the behaviour of teleplasm under the flashlight."[10]

Teleplasm held the promise of aligning the findings of psychical research with twentieth-century scientific ideas. Its unstable character, however, proved hard to classify. Typically, teleplasm issued from various orifices of the medium's body, whether the nose, mouth, eyes, or vagina. At times this condensation formed highly complex shapes, such as human faces, hair, hands, fingers, or intricately woven textiles. At other moments, it manifested as a pasty or "wooly" white mass resembling slim chords, rods, or membranes, its illuminated structure being visible to the eye as well as the camera.[11]

Indeed, the "elastic" or flexible nature of the substance seemed extraordinary to behold, for it remained malleable to the will of the discarnate intelligence.[12] Although it was understood as a distinctly spiritual substance, its physical attributes "could be observed and photographed for further study" and thereby measured and quantified in ways that other phenomena could not.[13] Frequently compared to a machine or apparatus, teleplasm was described as a piece of spiritual technology that "picks up voices from beyond the vibrations of the human senses."[14] It also retained the ability to be moulded into various forms, for it "simulates mechanical structures, simple or fabricated filaceous materials, as well as a whole range of 'living' organisms from the simplest to the most complex."[15] As Jenny O'Hara Pincock described, teleplasm "may be stepped down into the density of our material vibrations" to assume the shape and function of "vocal chords, hands, etc."[16]

In some respects, teleplasm was defined by what scholar Patricia Spyer refers to as "irreducible materiality" – marking the boundaries between object and investigator even while it served as the prime target of fascination, conjecture, and experimentation.[17] Akin to Evelyn Fox Keller's analysis of objectivity, investigators' conclusions positioned teleplasm as apart from and exterior to their scientific selves, who alone practised objective reasoning free of "desires, wishes and beliefs."[18] Yet teleplasm was not so easily categorized, its mutations and active engagement with the psychical researchers blurring the distinctions between subject and scientist.

In Boston the advent of teleplasm occurred in 1924. Early in the year, a series of organic illuminations appeared, including a sphere of light that represented the spirit Walter's eye, as well as "a luminous finger" that "showed the bones as in an X-ray photo, anatomically correct."[19] Walter eventually cooperated with J. Malcolm Bird and Hereward Carrington to obtain moulds of a teleplasmic hand in paraffin wax. After the investigators had prepared a bucket of hot wax before the seance, Walter proceeded to make a cast of the hand by placing its materialized form into the wax and then into cold water so that the mould would solidify. Walter then dissolved his teleplasmic hand, leaving only the cooling wax mould in its place. In this instance, the investigators did not actually witness the spirit hand materialize and be dipped into the hot wax. They consequently remained unaware of its composition, processes of development, and precise origin. To compensate, they fixated on the waxy results. Both Bird and Carrington carefully measured and documented the width, length, and diameter of both the hand and the wrist – comparing it to moulds of Margery's hand. Meticulously examining its features, from the fingernails to the position of the thumb, they noted anatomical differences between Margery's hand and the extremity that Walter had produced.[20]

These initial experiments generated a great deal of interest from North America to Britain but needed to be pushed further according to some. Acting as research officer for the Society of Psychical Research, Eric J. Dingwall came from London to Boston between December 1924 and January 1925 chiefly to investigate Margery and the teleplasm that emerged from her. He determined that he needed to see the actual substance materialize before his eyes, from its initial stage as "an amorphous mass" to its "formation of definite shapes."[21] In the hopes of capturing teleplasm on film, Dingwall solicited R.W. Conant of the Massachusetts Institute of Technology to operate a stereoscopic camera throughout the experiments. Fascinated by telekinesis and the independent voice of

Walter, Dingwall ambitiously sought to make visible "the actual mechanism of telekinesis."[22]

On January 6, 1925, Dingwall felt for the first time "a cold, viscous, clammy material" touch his fingers and hands, something that resembled "a cold damp tongue."[23] Portrayed as similar in texture to a "piece of cold raw beef or possibly piece of soft wet rubber," it became even more tangible when Dingwall used a luminous plaque to project its shadow onto the seance table. Dingwall described the silhouette as a "mass of black substance resembling about two inches of the end of a mittened hand."[24] Dingwall's depiction of its movements conjured a mental image of something wedged between a labouring life and a slow death. At one moment, the hand "flapped" against one of the sitters, and in another instance, Dingwall reported it "sliding across the surface of the table and then rose up against the luminous plaque."[25] Its methods of operation were carefully studied and recorded, the observation of the substance rendering it as distinct parts to pinpoint the origin of its force. Dingwall noted how the pressure it exerted emanated not from "the whole material ... but as if a harder object were covered with a clammy shell, and the pressure ... being exerted by it through the shell."[26]

During the seance on the next night, the mass appeared again, and Dingwall noted other distinguishing features. He reported once more being "struck by cool, clammy, apparently disk like object." Having taken hold of the hand, Dingwall observed, "the shape of the object was constantly changing. It appeared to lengthen and to widen, and occasionally parts appeared to be thickened, as if some internal mechanism was causing a swelling in parts of the mass."[27] Dingwall demonstrated his mastery over the substance by unveiling its mysterious mechanics through his eye and the flash of the camera. Capturing these strange structures as they emerged from the medium's mouth, nose, ears, and vagina rendered teleplasm tangible through mechanical, visual means. In one of the resulting images, the dark mass pouring from the medium's mouth spills out onto the table and slides toward the investigators, who intently watch its development and movements as it approaches them (see Figure 12). Albeit remarkable, the organic dynamism of the teleplasm revealed something that reached beyond the control of the experimenters. Political theorist Jane Bennet uses the phrase "thing-power" to describe an agent operating "in excess of ... human meanings, designs or purposes."[28] Similarly, the vibrant animation of the teleplasm as it most literally spilled from the medium and moved toward its investigators indicated that it was not passively waiting to be categorized. In contrast, the structure engaged the experimenters and the scientific project in a way wholly unexpected.

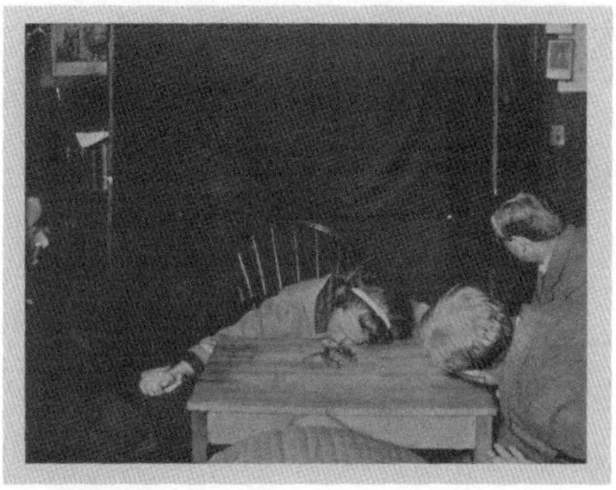

FIGURE 12 Margery seance, January 7, 1925. Photograph of Eric Dingwall's experiments with the medium "Margery" in Boston, Massachusetts. Dingwall sits beside the medium, holding her hand alongside other investigators. Margery rests face-first in a trance on the seance table with a black mass pouring from her mouth and moving towards the investigators.

Miraculous as these strange materializations seemed to be, Dingwall's study marked only the beginning of the appearance of teleplasmic structures in the Boston seances. Numerous investigators, from medical doctors to psychologists to entomologists to other scientists, continued experimenting with the teleplasmic materialization of Walter's hand and other organic forms in association with the Margery mediumship. Mark Richardson, E.E. Dudley, Hereward Carrington, R.J. Tillyard, Bracket K. Thorogood, and others joined forces and at times even competed with Le Roi Crandon and J. Malcolm Bird to observe, scrutinize, and photograph this phenomenon.[29]

Several of these photographs, particularly those documenting the appearance of Walter's hand, revealed anything but a sedate specimen. Psychical researchers reported at length their engagement with the hand, going so far as to obtain paraffin moulds and fingerprints, as well as to grasp and examine it, as Dingwall's experiments demonstrated.

This interactive teleplasm did not fail to draw interest, even when its authenticity was drawn into question. Controversies regarding the fingerprints and moulds arose in the early 1930s, beginning with psychical researcher E.E. Dudley's claim that the prints actually belonged to a living man – namely Mina Crandon's dentist, known only by his alias "Dr. Kerwin." Yet despite this exposure, fascination with the hand gained momentum. Innumerable reports and articles continued to be written on the subject well into the decade.[30]

Other sorts of teleplasmic forms also began to present themselves in the Boston seances to accompany the hand. One such form was what the spirit Walter referred to as his "talking head." It materialized almost immediately after Dingwall's investigations had ceased in February 1925, and as Bird described, it resembled a "small, lumpy, breathing, luminous" mass.[31] Appearing initially to the sitters as little more than a floating "vague sphere of luminosity," Walter promised that it would eventually be "a real head, bones and all features."[32] The luminous ball developed over the next few seances, "appearing like a vortex that would evolute and then involute ... rolling and unfolding, which suggested the kneading of dough." Meanwhile, it continued "learning how to breathe," according to Walter, beginning with a nasal hiss.[33] By the end of February, "a definite skull with jaws open" had become visible to the sitters.[34] By March, the sitters reported that when the head appeared again, they "could hear the bones of it hit the table."[35]

After it seemed apparent that the "talking head" experiments were endangering Margery's health, leading to problems with her "frontal sinuses," the experiments ceased.[36] Walter never again attempted to produce this exact type of materialization. Nonetheless, references to a teleplasmic head of another variety arose in the summer of the same year, this time in connection with the materialization of a "teleplasmic apparatus," which Walter claimed made his "vocal effects" possible.[37] Mixing two forms of teleplasmic material, Walter slowly constructed what he referred to as his "larynx" until it finally appeared fully "hardened" on August 13, 1925.[38] Upon turning on a red light, the sitters reported seeing on that evening a "mass of teleplasm overlying the entire face."[39] Connected to the medium's ear by a thin cord, it gathered first on her face and then shoulder, a much thicker cord running down the right side of her breast. Taken together, investigators' descriptions depicted the mass as a roughly formed trachea running up to a larynx. Reminiscent of the previous "talking head" manifestation, the "laryngeal mass" appeared to transform into a human face and head at select moments throughout the seance.[40] Mutating and vibrant, the teleplasmic mass was anything but inanimate.

Teleplasmic phenomena also emerged in Hamilton's Winnipeg seances at roughly the same time as they appeared in the Boston experiments. As early as 1921, Hamilton hypothesized that the telekinesis demonstrated through the medium Elizabeth Poole was the result of "psychic structures" that "appeared to be controlled by invisible operators manifesting intelligence of the human kind."[41] Theorizing that these structures at least "partially materialized ... so as to perform various actions," Hamilton obtained as much evidence as possible of such structures and the intelligences behind them.[42] From December 1923 to November 1924, Hamilton conducted several experiments with Poole similar to the wax mould experiments of Bird and Carrington in Boston. However, rather than an entire hand, Hamilton's initial experiments in this vein resulted in twenty wax moulds of fingers, toes, and their corresponding prints.[43] In addition to the actual wax forms, Hamilton also compiled lists of measurements, including each digit's length, width, and weight, as well as other physical characteristics of the skin and nails.[44] Although suspected to be moulds of the medium's control personalities, such as Robert Louis Stevenson and William T. Stead, no definitive proof could be obtained to verify this conjecture at the time. Recalling these experiments later, however, Hamilton asserted that their meaning changed "in light of the greater materializations to come," pointing more definitively to the survival of personality.[45] Indeed, these earlier materializations seemed to belong to "a whole," indicating the "first steps" in a series of experiments that followed with the introduction of the medium Mary Marshall.[46]

By the late 1920s, Poole's gifts appeared to be lessening. Although she still continued to play a role in the seances, the newly introduced medium, Mary Marshall, gained pre-eminence in Hamilton's experiments. She, like Poole, had demonstrated supernormal gifts before entering Hamilton's psychical laboratory, as described in Chapter 2. Consequently, Hamilton and his fellow investigators expected her inclusion to yield a great deal, which, according to the reports, initially led to disappointing results.[47] However, she soon demonstrated a profound ability to produce an ample amount of teleplasm. On June 4, 1928, teleplasm first became visible to the cameras' lenses. The main intent of the experiment was to photograph the telekinetic phenomenon induced by Walter. However, when Hamilton examined the negative under magnification, he noticed slim, white cords flowing from Mary Marshall.[48] Fixated by the substance, Hamilton altered the direction of his experiments so that he might analyze it. Although still keenly interested in telekinesis, trançe, and other phenomena, he fastened most of his experimental goals upon the studious examination and further development of teleplasmic materialization.

Hamilton was not to be disappointed. At first the teleplasm took the shape of cords or loosely organized masses, but within the following year it had formed into "capsules," in which certain identifiable faces could be seen.[49] The likeness of William T. Stead appeared first – a personality that had manifested in some fashion during Hamilton's experiments since 1921. This event was followed by the appearance of many other well-known cultural and literary figures, including Charles Hadden Spurgeon, Robert Louis Stevenson, William E. Gladstone, and a host of others.[50]

In addition to faces, teleplasm transmuted into several different forms. Sometimes it retained its initial disorganized, mass-like structure. At other moments, the ghostly condensation became more complex, taking on the likeness of hair, headdresses, and beaded necklaces. Teleplasm continued to shift into more varied and intricate shapes as Hamilton carried on with his experiments into the 1930s. Transforming into brilliant garments, veils, handkerchiefs, and model ships, as well as various body parts such as hands, fingers, and truncated torsos, teleplasm evaded classification.[51]

At one moment a biological organism and at another a complex machine, teleplasm took on as many uses as it did forms. Asserting that teleplasmic structures served a primarily functional role, Hamilton surmised that the spirits used it to "convey energy from the body of the medium to move inanimate objects at a distance," to provide the structural basis of "talking machines" through which the spirits could speak, and finally "to take on the appearance of a part of the whole of the physical body."[52] The procedures and language the spirits used emphasized the instrumental quality of teleplasm.

The spirit of James H. Hyslop, a psychologist and prominent member of the American Society for Psychical Research (ASPR) until his death in 1920, described this characteristic of teleplasm in a Winnipeg seance of February 1932. Speaking through Harry A.V. Green, the ghost of Hyslop described how discarnate intelligences "operate directly upon [teleplasm], it being pliable to thought."[53] Comparing teleplasm to "any other matter," he explained that the spirits "subjected it to forces" and that the main difference between teleplasm and other physical substances was that "its pliability is in the inverse ratio to its relationship to your time ... If I were able to take it out of your time it would be so easy, but the stability of matter interferes with its pliability to thought."[54] It was this material quality that made teleplasm such a useful substance with which "to make things visible."[55] Reportedly "greasing the wheels" before a coming materialization of a "wax apparatus," the spirit of Walter spoke of teleplasmic structures as instrumental "struts," "mechanical contrivances," or "machines" that he used to build apparatuses or perform certain functions.[56]

However mechanical and pliable, teleplasm took on a life and will that seemed entirely its own. As in the Boston seances with Margery, the Winnipeg Walter frequently reported that he was building apparatuses beyond the seance participants' immediate perceptions, such as his

FIGURE 13 "Voice Box," 1929. A photograph of the spirit Walter's "talking machine," taken in the Hamiltons' psychical laboratory in Winnipeg, Manitoba. Shaped like a "tiny trumpet" and "very hairy," the black mass pours out from the left ear of the medium Mary Marshall while she sits in the open cabinet in trance.

"talking machine" – the teleplasmic device through which he spoke.[57] The machine appeared on July 10, 1929, resembling a "tiny trumpet" as it poured forth from the medium's ear. With a "husky mannish voice," the talking machine instructed Hamilton to fire the cameras so that it could be captured on film; this was followed by a sucking noise as it contracted back into the mediums body.[58] Like many other teleplasmic technologies, it appeared biological and even "very hairy," conveying its intimate relationship to the biological processes of the medium (see Figure 13).[59]

The organic appearance of this teleplasmic structure and others was not at all lost on Hamilton, who noted how the substance "possesses many features of great interest and significance from the standpoint of biology."[60] Whereas its more organized incarnations formed into "fingers, hands, arms or faces" whose "self-luminous" quality paralleled "the light of the glow worm," its more "unorganized" varieties resembled "fine threads, rods, thin membranes."[61] Moving and acting in the seance room, even directing the experiments at times, teleplasmic structures appeared as vitalized agents with a force that could not be easily contained or controlled. Skewing boundaries between the spiritual and the material, the organic and the mechanical, teleplasm's vigorous animation demonstrated its propensity to engage the investigators and their experiments on its own terms. Latour defines an "actant" as something that "modif[ies] a state of affairs" to "allow, afford, encourage, permit, suggest, influence, block, render possible, forbid."[62] Similarly, the teleplasmic structure shaped the experiments of the psychical investigators, defining what they could do and what they could not. Despite promising to affirm the place of psychical research in twentieth-century scientific pursuits, it skewed the boundary between investigator and compliant subject, shifting the authority in the seance room. However, teleplasm would play only one part in redefining the empirical project within the psychical laboratory. The mysterious substance was joined by other ghostly entities who continued to distort the distinctions between the experimenters and that upon which they experimented.

Experimental Apparitions: Discarnate Scientists at Work in Their Laboratories

T. Glen Hamilton explained in an unpublished article how he knew when to prepare for the coming of certain phenomena: "The answer, incredible as it may seem, is that it is the trance beings who supply the needed information ... It is they who plan and direct often during a series of sittings extended over many weeks, all the more important procedures."[63] Hamilton knew that by relinquishing control over the seance room, his experiments could be brought into question. Nevertheless, he insisted that such conditions were necessary for the success of psychical experiments. In fact, by this act of cooperation with metaphysical powers, his experiments not only proved the "reality" but also the "scientific acumen of our invisible collaborators."[64]

Unruly and exotic ghosts proliferated in the seance room, as revealed in Chapter 3. Yet the spirits to whom investigators granted the most esteem in the seance room reflected their dedication to empiricism. In fact, these discarnate entities were in many ways all that the psychical researchers perceived themselves to be, encapsulating what Donna Haraway describes as the "self-invisible, transparent" character of the "modest witness" that remained at the heart of the scientific enterprise. Most literally, since ghostly scientists were "not ... polluted by the body," their ability to perform experiments granted them extraordinary power in the seance room that, at moments, usurped the authority of their earthly counterparts.[65]

Organizing and carrying out experiments with the aid of "many wonderful workers," discarnate scientists diligently laboured in their spiritual laboratories to empirically test the limits between this world and the next.[66] Hamilton explained that in his psychical laboratory, all the spirits contributed in some fashion to the success of the experiments. Several spirits acted as "technicians" of the experiments. Nevertheless, they did so only "under the direction of many scientists."[67] Thus, much like in the more earthly realm, everyone may have played a role, but only a few – mostly masculine – spirits took charge, determined to "discov[er] the laws governing communications between the two states of existence."[68]

These authoritative masculine spectres were by and large dead psychical scientists who shared the commitment to empiricism of their more earthly counterparts. The presence of Camille Flammarion, a deceased French astronomer, was instrumental to the Winnipeg experiments on several occasions. Famously known as one of the first to scientifically pursue the paranormal, Flammarion initially maintained close ties with the French movement of Spiritism, founded by Allan Kardec in the mid-nineteenth century. Eventually, in the 1860s and 1870s, his empirical ambitions alienated him from the movement. Despite the resistance he received, Flammarion persisted in his psychical experiments well into the twentieth century. Flammarion was one of ten who comprised the first committee of the Institut Métaphysique International in 1919 – an affiliation he maintained until his death in 1925.[69]

The spirit of Flammarion emerged from the medium Elizabeth Poole in the Winnipeg seances fairly soon after his passing. Poole disliked this spirit who controlled her, referring to Flammarion derisively as "that foreigner!"[70] He nevertheless used her to cooperate with Hamilton and others, a collaboration that resulted in various phenomena. Hamilton entitled the most impressive Winnipeg experiment associated with Flammarion as "The Handwriting on the Wall." On June 10, 1931, Walter instructed that one camera be set up to capture "a name on the back of the cabinet." Upon obeying Walter's orders, Hamilton developed the photograph to reveal the name "FLAMMARION" written in teleplasm at the back of the cabinet as Walter had indicated.[71]

The spirit of the famous British scientist and psychical researcher William Crookes also reportedly orchestrated many of the experiments of the Winnipeg seances, mostly in association with the apparitions of Katie King. Similar to Flammarion, Crookes had been a highly regarded psychical researcher in life. An avid chemist and physicist credited with the discovery of thallium and the invention of the radiometer, Crookes applied the methods of natural science to his various paranormal investigations.[72] As the previous chapter outlines, one of Crookes's most famous investigations was with the medium Florence Cook, whose seances resulted in the brilliant appearances of the spirit Katie King.

When Katie King made herself known from 1930 onward, Hamilton immediately connected her presence to Crookes and his earlier experiments.[73] Based on the visual evidence, as well as the testimony of the spirits to whom Hamilton granted "the status of truth-telling witnesses," the appearance of Katie King in the Winnipeg seance was a wondrous repetition of Crookes's experiments.[74] Crookes, who had died in 1919, did not make himself explicitly known until almost a year after Katie resurfaced, in January 1931.[75] The close association between these two spirits nevertheless prevailed, especially when Crookes began speaking through Susan Marshall – the same medium through whom Katie King emerged. Closely resembling his more earthly persona, the ghost of Crookes reportedly inspected the cameras, built "glass tubes" presumably similar to the electrical discharge tubes he had invented in life, and ensured that the unseen technology was properly in order.[76] Adjusting "cords in the cabinet" and employing ethereal devices referred to as "colored globes" and "shells," the spirit of Crookes diligently worked so that Katie King's veil, face, and body might appear.[77]

The importance of discarnate scientists such as Flammarion and Crookes was demonstrated by the experiments they conducted, as well as the level of credibility they afforded the paranormal investigations more generally. As they were recognizable characters who had contributed to science in years past, their performance in the seance affirmed the methods of the living men who conducted such paranormal investigations. By reflecting a vision of what Daston and Galison refer to as the "scientific self," these spirits mirrored the human psychical scientists with whom they reportedly worked.[78] Their presence in the seance room reinforced the conviction that Hamilton and others were continuing the investigative work of these scientists, building upon their accomplishments to draw psychical research into the twentieth century.

That being said, the spirit chiefly responsible for the scientific procedure of the seance room in not only Winnipeg but also Boston was no scientist at all, at least not in life. Walter, the ghost of Margery's deceased brother, did not have the credentials of Flammarion or Crookes. Walter was a working-class man before his untimely death, and his education had been decidedly minimal, scientific or otherwise.[79] It was only in death, apparently, that he gained an appreciation for and skill with the scientific method. Hamilton only very cautiously accepted Walter due to the ghost's precocious working-class antics. Yet this spirit's aptitude in directing successful experiments led Hamilton to believe that Walter's dedication to scientific empiricism paralleled his own.[80] Like Hamilton, Walter insisted upon strict methods and the use of apparatuses, demonstrating his mutual faith in the ability of technology to uncover the newfound realities of the twentieth century. Walter, in fact, was instrumental to the introduction of the bell-box in Hamilton's psychical laboratory. Reportedly, Hamilton began such experiments only at the request of Walter, who insisted that such an instrument be built and introduced into Hamilton's psychical laboratory in March 1928.[81]

Walter continued to encourage more elaborate tests in the years that followed, proving over and again his invaluable role in the Winnipeg experiments. Consequently, Hamilton described Walter and those who worked alongside him "as men working in a laboratory – and not as ghosts or spirits."[82] Emphasizing the masculine and scientific nature of his endeavours, Hamilton insisted that Walter and other spirits who emerged "appeared to be men who knew what they could do, said so, and succeeded in carrying out their stated program with dispatch and accuracy."[83]

Such devotion to the scientific method correlated the Winnipeg Walter with the main spirit of Le Roi Crandon's Boston investigations. In contrast to Margery's assumed scientific ignorance, the Boston Walter proved astute at scientific terminology and technical skill, demonstrating knowledge of twentieth-century physics and other natural sciences.[84] Similar to the human investigators with whom he worked, Walter apparently set up a variety of "super-physical apparatus," which he carefully monitored and employed depending on the phenomenon he wished to produce.[85] And like his Winnipeg double, the Boston Walter experimented alongside the human psychical scientists, carrying out specific programs to produce materializations and telepathic or telekinetic feats. As J. Malcolm Bird reported, Walter described himself as "a scientist, faithful unto the cabinet and scales."[86]

Walter's demonstrated scientific talent led the psychical researchers of both Winnipeg and Boston to relinquish more and more control to him. As a result, he gained a level of authority that at the very least equalled that of the psychical researchers. All tests, methods, and personnel, Crandon maintained, had to first be approved by Walter – a condition to which he insisted each participant agree in writing.[87] In this capacity, Walter was also given the power to cut short experiments when he felt it necessary, most often for the reason of the medium's mental and physical health.[88]

In Winnipeg, Hamilton permitted Walter an extraordinary degree of freedom to coordinate experiments and dictate how the medium should be handled during the seances. Walter himself demanded this level of authority. On October 29, 1929, Walter explained, "I know very well that you want everything above board and it shall be so, but you must leave me to do it in my own way, not your own."[89] He similarly insisted on dictating how the medium was to be treated during the seance. Walter was adamant that no one was "to move or touch the medium unless I give orders ... don't touch her in this room, understand, without my permission."[90] As though offering an example for the human participants to follow, the other spirits made clear that they too would surrender their own power, "as Walter wanted the experiments to be successful; to achieve this he alone could control."[91]

Witnessing and experiencing the presence of these ghostly scientists, psychical researchers conceived of the identity that they believed they shared with these spirits as extending through time and space in order to make their claims to scientific authority universal and enduring. Avowing the same principles of scientific and technical precision, investigators believed they had somehow met a very literal kindred spirit – a "*vir modestus*" much like themselves who, with diminished bodies and heightened minds, sought to investigate the realms of the natural world and beyond.[92] So convincing were the spirits in this role that discarnate scientists gained a profound degree of power within the seance room. Dictating the methods of handling the medium, the timing of each experiment, and the devices to be employed, spirits such as Flammarion, Crookes, and Walter transformed the terms of psychical science to usurp the authority of their earthly investigators.

The Medium as Machine: Technologies of the (Dis)embodied

Teleplasm and the ghostly scientists engaged the scientific project of the psychical laboratory with unique propensities and tactics. They were, at the same time, fundamentally dependent on the clairvoyant. Discarnate scientists could operate within the seance only through the body and mind of the medium. Meanwhile, psychical scientists emphasized how teleplasm maintained an intimate "relationship to the basic life energies of the human body," which "appeared to be connected with the celiac and abdominal plexus of the medium."[93] The medium may have been constructed as a passive scientific subject, yet the gifts of the clairvoyant also made psychical research possible. Characterized as a delicate machine through which phenomena emerged, the medium played a critical role in the scientific project as it was conceived in the interwar psychical laboratory.

From the late nineteenth century onward, the ordering of the seance was based upon the composition of a battery, in which positive and negative forces emanating from the sitters were aligned to bridge the metaphysical gap between this world and the next.[94] Hamilton described how sitters and medium together composed "a machine ... functioning most efficiently when in full working order."[95] Or, as the spirit of Lucy through the medium Susan Marshall succinctly put it, "We are all cogs in the wheel."[96] Spirits arranged their experiments accordingly, drawing "wires" from the sitters to the medium in order to procure the energy required.[97] By and large, however, the metaphor of "machine," battery, or instrument was most consistently applied to the medium. According to media and communication theorist Jeffrey Sconce, the nineteenth-century medium was viewed as "a complex receiver who channelled the mysteries of spiritual electricity."[98] This characterization of mediumship continued well into the twentieth century. In a 1936 letter to Jenny O'Hara Pincock, Fidella Dinsmore Papa of Chicago explained the medium William Cartheuser's telepathic ability in precisely these terms, but in a way that reflected twentieth-century advances in telecommunication. As Papa made clear, "psychic waves of thought" were "a phenomenon analogous in every way to radiophonic communications, with electric waves travelling in spherical form." These waves were "perceptible and can be caught everywhere with the sole condition that the receiving apparatus be regulated on the same length of wave." Cartheuser acted as the "receiving apparatus" in this scenario.[99] Referred to at various moments as an "instrument," "battery," or "human radio," Cartheuser served as the means through which the spirits drew the necessary "chemical force" for the production of manifestations and direct voice.[100] Thus successful spirit communication depended on the "instrument's vibrations," which could be put "badly out of order" due to bad health, mental unrest, environmental conditions, or the negative attitudes of the sitters.[101]

Both psychical researcher and spirit required precision and technical skill to operate upon this complex "human instrument." Cartheuser's main spirit, Dr. Anderson, described how the spirits systematically examined the medium to see whether he or she was "equipped within." This would ensure the presence of "enough chemical force for our purposes." If the medium was found to have satisfactory levels of the necessary compound, the spirits extracted "a substance like smoke or fog" from which they "assembl[ed] ... into a more material substance." By means of this now-more-solidified material, the spirits vocally and physically interacted with sitters in the seance room.[102]

The manner in which earthly and spiritual scientists described the function of mediums only further confirmed their instrumentality to the success of each experiment. The spirit Dr. Anderson maintained that the "audibility of the voice is created by re-using the stored up sound waves you create by your singing and conversation, very much like a storage battery."[103] Although this energy could be potentially stored anywhere in the seance room or cabinet, its primary container seemed to be the medium. Hamilton was quoted as saying, "Our mechanical devices are dependent upon the stored energy of inanimate nature. In telekinesis ... the force exhibited depends upon energy stored in the body of the medium."[104] Rather than merely an empty box, the medium's body acted as a machine that extracted energy from its environment and restructured it to produce paranormal phenomena. As Hamilton reported upon examining the telekinetic powers of Elizabeth Poole, "the medium's organism appeared to be a reservoir and a transformer of energy drawn from its surroundings."[105]

With the successful production of ectoplasm, the medium's role as a dynamic mechanism for the storage and conversion of paranormal power seemed even more pronounced. The spirit of Walter insinuated that he used the medium to store the teleplasmic substance, particularly in its incubatory stage. Like "grafting skin from place to place," Walter moved the ectoplasm directly from the fourth dimension to the medium's body so that it could develop and be ready for full materialization.[106] "I have her brain filled with ectoplasm, which I could produce," Walter claimed of the medium Mary Marshall during a seance in April 1929.[107]

Judith Butler describes how the feminine is represented as a mere "receptacle" rather than an autonomous agent.[108] Female psychics may have paralleled this description, yet they regularly evaded such categorization as well. An organic locus through which paranormal forces emerged, the medium enabled psychic energies to manifest, contributing to the conviction that the seance was "a technological event itself," to use the words of English and media scholar Jill Galvan, "one centred on the medium."[109] Arguably, the investigators' fixation on the medium's body granted this so-called passive subject a great deal of influence within the seance room.

The power that mediums exercised emerged from the conviction that they were, as Sconce describes, "wholly realized cybernetic beings – electromagnetic devices bridging flesh and spirit, body and machine, material reality and electronic space."[110] The characterization of mediums as machines placed them firmly in line with the twentieth-century belief in technoscience, much as Chapter 5 outlines. Of all the technology in the psychical laboratory, the medium proved to be absolutely essential to paranormal experiments. Viewed as machines with extraordinary potential to connect beings across time and space, the psychic's electrical-equivalent "nerve force" provided the energy required to produce telekinetic and teleplasmic phenomena.[111] Claiming this role for themselves, mediums gained incredible power, even when restrained and monitored. Transgressing the bounds of passive mediumship, their powers combined with spirits and teleplasm to engage and refashion the empirical practice of the interwar seance room, while subverting the gendered and sexual hierarchies at work.

A Scandalous "Thing": Vaginas, Rods, and Mystifying Sex

Ozora Davis, president of the Chicago Theological Seminary, attended a 1927 seance with Margery. Although given free rein to bind and restrain the medium as he deemed necessary to ensure that "the control was perfect," Margery's restless movements and moaning proved "disturbing" to both him and his wife, who had accompanied him.[112] The reasons for their discomfort were left unspecified in the record. Nevertheless, investigators and sitters frequently noted the erratic cries and movements of Margery, especially during the production of teleplasm – Margery's groans and grunts being compared "with those of a woman in labour."[113]

The production of teleplasm maintained an intimate connection to the medium's body, specifically her reproductive and sexual faculties. In large part, this link was established due to the precise point of origin on the medium's body where the teleplasm emerged. Teleplasm, psychical researchers reported, typically poured forth from various orifices of the medium's body, including nose, ears, and mouth.[114] In Margery's case, however, the most common orifice through which the teleplasm appeared was her vagina. Climbing out from between her legs to perform various experiments, strange teleplasmic structures, resembling either a hand or what became known as Walter's "terminal" or "rod," continued to emerge from Margery's vagina in the years that followed – so much so that the occurrence dictated Margery's standard of dress and position in the seance. Margery was urged to wear a robe and stockings into the seance room, with the understanding "that under the cover of darkness she [would throw] the robe more or less open, with the theory that this facilitates the necessary projections from her body into the theatre of action."[115] Several of the investigators who witnessed this teleplasmic structure noted that it seemed to be connected to the medium by a slim projection resembling "an umbilical cord," which yet again correlated the projection of teleplasm to a birth.[116]

Margery and other mediums actively engaged the researchers who examined them during experiments with teleplasm. Psychics guided the hands of their investigators over their "head, face, neck, breast and underarms." Since they were understood as "entranced" at the time, it was claimed to be an act by the spirits to ensure that the medium was thoroughly inspected.[117] Conveniently, it also ensured that the respectability of the medium and psychical researcher remained unscathed. The potential subversive nature of the act was nevertheless heightened when the psychic suggestively groaned in response to the touches of her human investigators, an incident that often occurred when the researchers examined teleplasmic protrusions.[118] Investigators endeavoured to systemize and discipline the perception of touch in the seance, as Chapter 4 describes. But troubling encounters between medium, teleplasm, and investigator reignited the transgressive nature of this perception.

Dingwall was the first to experience the vaginal psychical discharge during his experiments with Margery and Le Roi Crandon in early 1925 – an encounter he was able to subsequently capture on the photographic plate (see Figure 14).

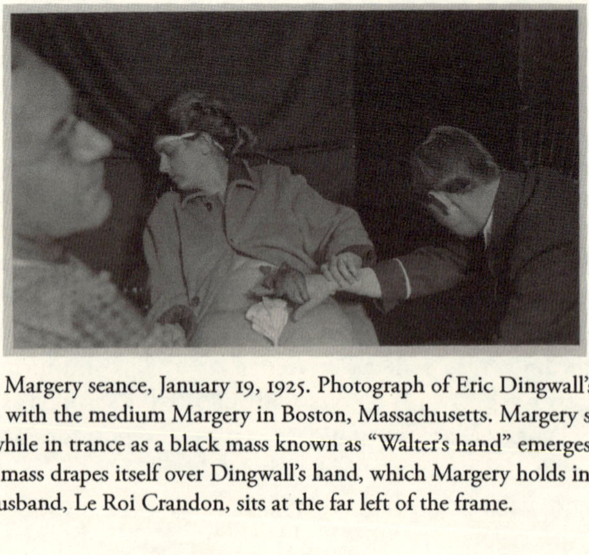

FIGURE 14 Margery seance, January 19, 1925. Photograph of Eric Dingwall's experiments with the medium Margery in Boston, Massachusetts. Margery sits and leans back while in trance as a black mass known as "Walter's hand" emerges from her vagina. The mass drapes itself over Dingwall's hand, which Margery holds in place. Margery's husband, Le Roi Crandon, sits at the far left of the frame.

Trained as an anthropologist and being a popular writer of sexology, Dingwall may in fact have harboured interests that inspired the disconcerting eruption from between Margery's legs.[119] His particular fascination with sexual functioning deemed abnormal, or "queer," reflected an explosion in the field of sexology during this period. Despite recognizing the fluidity of sex, North American and British sexologists also constructed a set of normals that could be defined only in contrast to so-called aberrations. This method of analysis became especially pronounced in the interwar period when sex and gender were mapped upon the body in new ways through advancing genetic, chromosomal, and hormone research.[120] The teleplasm emerging from Margery's vagina no doubt struck Dingwall as a sexual deviation of a unique kind. During Dingwall's experiments with Margery's teleplasm in 1925, he described how he was guided to feel the "soft, clammy and cold" substance by the medium. He pressed his finger up against the connecting structure flowing out from Margery to the fully formed teleplasmic object. As Dingwall's fingers roamed titillating close to the medium's most intimate regions, "the Psychic groaned each time."[121]

This suggestive interaction was the first of many to follow, including during a 1926 investigation by a commission of inquiry sponsored by the ASPR. The investigators included psychologist Henry Clay McComas, in addition to two other men from John Hopkins University: Knight Dunlap, a professor of experimental psychology; and R.W. Wood, a physicist. The commission aimed to conduct an impartial investigation by expert personnel to ascertain "what phenomena occur and the conditions under which they occur."[122] Yet in the end, such ambitions proved difficult to fulfil, as the commission disintegrated most literally over the body of the medium.

The commission began with an "entirely informal" sitting on January 26, 1926, leading to a few standard rings of the bell-box but little else.[123] The first official sitting occurred on the following night. At this time, they tested Walter's ability to identify letters in the dark and witnessed him levitate objects such as a basket, a megaphone, and a luminous disk.[124] Teleplasm, however, would not emerge until the next official sitting. According to the records of January 28, "a mass giving a silhouette" appeared, accompanied by "groans suggesting labour pains" on the part of the medium. When investigators questioned Walter as to why teleplasm materialized only during the second official sitting, the spirit replied that he had been conducting "a little psychological experiment to see how these men would behave. Wood during the whole sitting kept pawing about over the abdomen and thighs of the Psychic. I did not dare bring out the teleplasm under these conditions."[125]

It was not until the third seance of the committee on January 30 that the investigators gained the opportunity to interact with the teleplasm as it emerged from the medium. At precisely 10:30 p.m., McComas reported that he felt something "cold and moist" slap him on the hand, which was followed by Dunlap reporting the same. In the next few minutes, both investigators described something "cold and moist" and "rib-like" touching their hands, while the psychic "twist[ed] and groan[ed]." Within a few minutes, the medium grabbed the hands of both McComas and Dunlap in turn, letting them feel the "cold, smooth, cylindrical something." McComas described it as being "like a cylindrical substance about ¾ inches in diameter with a hard, bone-like center or axis." From this point forward, the investigators referred to the substance as the "thing." Meanwhile, their physical examination of the medium became increasingly bold. Wood claimed that he "squeezed" the teleplasmic rod "very hard." McComas reported "feeling Psychic from the thigh to knee," and Dunlap related that "the Psychic has silk bloomers on." At the same time, the medium repeatedly drew the hands of the investigators to press against the teleplasmic rod, as well as her lower abdomen, where McComas reported he felt something like a "belt."[126]

This seance resulted in a huge controversy, particularly over Wood's reported hard "squeeze" or pinch, which Crandon insisted defied the agreement between him and the members of the ASPR commission. This agreement specified, "No psychic structure shall be touched or interfered with in any way except with the consent of the control (Walter)."[127] Emphasizing both the sensitivity of the teleplasm, as well as its intimate connection to the medium's sexual functions, Crandon reported how the medium was struck by a case of "nausea and vomiting," as well as severe "catamenia," or excessive menstrual bleeding, as a consequence of Wood's rough handling.[128]

An even more troubling incident occurred on this night. Although it did not appear in the official records, Crandon reported the upsetting event in a letter to Daniel Day Walton, the trustee of the ASPR counsel. Apparently, Dunlap had stated during the January 30 seance that "the whatcha-may-call-it feels to me like a——." Crandon left the exact wording ambiguous, merely stating that Dunlap's phrasing was highly objectionable: "There were plenty of similes which this man could have used in the presence of the female stenographer, to say nothing of the decent respect for the Psychic, even though she was unconscious in trance."[129] Such actions and language, Crandon argued, constituted "behaviour impossible for a gentleman under the circumstances."[130]

As the exact wording is apparently lost, one can only conjecture what Dunlap may have compared the teleplasmic rod of Walter to. However, it is not difficult to imagine how a "cylindrical substance" resembling "a rigid rod" and emerging from the genitalia of the medium may have provoked those in the seance room. That the teleplasm seemed intimately connected to the sexual functions of the medium, and that it represented a teleplasmic "something" of a male intelligence, only multiplied the possible correlations that Dunlap could have made.[131] Walter's "terminal" may have been acclaimed as extraordinary – the "teleplastic telekinetics" that it "exemplified" being "hithero unrecorded in the history of psychical research."[132] Yet its point of origin, phallic aesthetic, and feel all pointed to the real possibility that Walter was pouring forth from Margery in ways that fundamentally altered not just her mind but also her sexed body.

Butler identifies the phallus "as the privileged and generative signifier."[133] Although having a "phallus" is commonly perceived as occupying "the masculine position within a heterosexual matrix," Butler challenges this view, contending that the phallus is much more "ambivalent" and uncertain.[134] Linking her analysis to lesbian sexuality or the "lesbian phallus," Butler argues that the "plasticity of the phallus" is exemplified by its placement in association with the lesbian and thus creates a case in which "the signifier is significantly split, for it both recalls and displaces the masculinism by which it is impelled."[135]

Walter's teleplasmic terminal, jutting forth from Margery's vagina, displaced the phallus from its rigid masculine position. Medium, spirit, and teleplasm thus undermined the phallus's ability to signify power by revealing its fundamental "variation and plasticity."[136] In response to being examined by psychical scientists who aimed to penetrate the metaphysical mysteries of her body, Margery, through spirit and teleplasm, produced the material symbol of masculine prowess and penetration itself. Engaging the science of the paranormal in ways entirely unexpected, Margery exceeded the limits of her own gendered body and perhaps something more. By mimicking the masculinity of her investigators, she drew attention to their own corporeality. Psychical researchers had worked hard to present themselves as objective scientists untarnished by their "biasing embodiment."[137] Yet faced with the symbol of their masculine position of dominance projecting from the medium, they witnessed the subversion of their claims to authority. Displaced and distorted as it emerged from the medium, the phallus transformed from a site of power to something that hearkened to the investigator's inability to move beyond his subjective body. No longer distinguished from the subjects that he studied, the psychical scientist found himself entangled in the carnality of spirit, teleplasm, and even the medium herself.

***

Teleplasm, spirit scientists, and mediums were all praised for their unique contributions to paranormal experiments, which propelled the efforts of psychical researchers into the twentieth century. Teleplasm was reconceptualized along the lines of atomic energy. Spirit scientists affirmed psychical researchers in their project to carry paranormal experiments into a new era. Psychics were envisioned as complex machines who had the potential to unravel the mysteries of time and space. Each in his or her own way held the promise of aligning psychical research with twentieth-century scientific ideas.

Yet their engagement with their investigators ultimately transformed the terms of empirical enquiry in the seance, destabilizing the distinctions between material and spiritual, mechanical and biological, scientist and subject. Entering the fray of the scientific project as it was conceived in the interwar seance, teleplasm, spirit scientist, and medium blurred the boundaries between investigator and subject. Teleplasm's multiple mutations and dynamism made it instrumental to experimental results, as well as an active agent that participated in and reformulated experiments for its own ends. Defying easy classification and control, it slipped from the object of study to something that reshaped and even directed investigations initially intended to classify and control it. Spirit scientists practised alongside their earthly counterparts, building and manipulating various apparatuses in their own realm, while encouraging psychical researchers to do the same. Gradually, the power that these spirits gained to orchestrate tests and dictate procedures made it difficult to discern who was experimenting with whom. Meanwhile, mediums seemed to take on an identity that strikingly resembled what Haraway describes as a "cybernetic organism, a hybrid of machine and organism." When teleplasm poured from their most intimate parts, psychics evaded a "fixed location in a reified body," disrupting and even dismantling masculine authority.[138] As male investigators watched and examined the symbolic site of their sexual power jutting out from the female medium, they found themselves demoted in their position as objective observers, as they were unable to escape the bounds of the body. Their exclusive hold on the scientific project in their psychical laboratories could no longer be sustained, as their unstable subjects reformulated what it meant to be scientific in the seance room.

The Knot Unravelled: An Epilogue

This animistic stage in primitive man ... none of us has passed through it without preserving certain residues and traces.
– Sigmund Freud, "The Uncanny"[1]

In the years following the First World War, society did not reject the existence of the paranormal. Instead, psychical phenomena functioned as an articulation of the culture itself. This study has examined a specific generation of psychical researchers who jointly experimented and corresponded with one another from Canada to the United States to Britain between 1918 and 1939. Sharing data, methodology, technology, mediums, and even spirits with one another, they legitimized their work while demonstrating the intricate transnational lines of communication that formed over the pursuit of the paranormal. Far from simply an extension of Victorian spiritualism, what these investigators endeavoured to create was a science fully rooted in the twentieth century. At times upholding and at other moments utterly subverting broader cultural perceptions of gender, race, and class, their enactment of science called into question who was to have power over whom. Embodied performances played a fundamental role in these reconfigurations, affirming, contesting, and at times disassembling claims to empirical authority. Discontent with simply reiterating scientific discourse and technological praxis, psychical researchers, mediums, and spirits ultimately transformed the tenets of empiricism in the seance-turned-laboratory, while offering new possibilities by which gender, class, and racial power could be subverted and remade.

By the end of the 1930s, many of the people who made up this network had died or moved on, leading to its gradual disintegration. As early as the mid-1930s, both T. Glen Hamilton and Walter Franklin Prince had passed away.[2] By the 1930s J. Malcolm Bird had distanced himself from the Margery case and psychical research more broadly, including the American Society for Psychical Research, leaving his preparations for a third volume of the Margery mediumship to his successors at the society, who published his and other records in 1933.[3] Henry Clay McComas had also come to the end of his career by the mid-1930s, his 1935 account of his experiments with several mediums, including William Cartheuser and Margery, being the last book he ever published.[4] Jenny O'Hara Pincock dissociated herself from the medium William Cartheuser from 1935 onward. Believing that he had succumbed to the temptation of profits over and above spiritual revelation, she felt it necessary to break off their relationship. Although she still believed in the merits of investigating the paranormal, she had cut many of her associations with the Church of Divine Revelation and the Radiant Healing Centre by the end of the decade.[5]

Le Roi Crandon died in 1940, to be followed by his wife, Mina Stinson Crandon, a year later on November 1, 1941. The once-famous Margery had lost the battle with her detractors and found herself discredited by the time of her death. Unfortunate circumstances, however, would not leave her completely despondent, her dying words revealing a wilful, defiant woman. Upon being pressed to reveal her secrets regarding how she produced the phenomena she did, Margery reportedly retorted, "All you psychic investigators can go to hell."[6]

This account is not intended to suggest that psychical research came to an end by the 1940s. As others historians have documented, psychologists Joseph B. Rhine and Louisa E. Rhine had already begun to reshape the field by this time. Establishing a laboratory at Duke University in 1935, the Rhines dedicated themselves to experiments in statistically quantifiable instances of thought transference. Dissatisfied with their experiences with Margery and others, they distanced themselves from what had come before by insisting upon distinct nomenclature for their work. Replacing psychical research and telepathy with "parapsychology" and experiments in "extrasensory perception," their efforts marked an influential shift in not merely terminology but methodology and focus as well.[7]

However, some took a less radical direction, including members of the network that this study has examined. Despite having parted ways with O'Hara Pincock, Cartheuser continued holding seances throughout the

United States and Canada well into the 1940s.[8] Hereward Carrington carried on with his experiments in psychical research while the Second World War was raging and beyond. Despite complaining that the war limited his access to the proper equipment and funds, he insisted that his field contributed to the breakdown of mechanistic thinking in science and would eventually prove to be "the most influential and important" of all the sciences.[9]

Lillian Hamilton continued to investigate well after her husband had passed away. By the 1940s, the mediums Elizabeth Poole and Susan Marshall had died. Harry A.V. Green also discontinued acting as a medium, although he still endorsed the Winnipeg experiments until his death in 1942.[10] Mary Marshall acted as a medium for Lillian Hamilton's experiments for many years afterward and was joined by others, including a woman known as "Mrs. Wither," whom Walter gave the name "Beulah." In these experiments, Walter continued to build teleplasmic apparatuses and forms, including a miniature face of the late T. Glen Hamilton in May 1939.[11] In 1943 Hamilton revealed himself again, this time in the form of several trance scripts written through the medium Mary Marshall. In these letters, addressed to Lillian, he seemed to have retained much of his earthly knowledge, but his purpose in joining the experiments had changed. Rather than fulfilling a strictly scientific purpose, which he had advocated so strongly in life, Hamilton's spirit conveyed that his "object in writing these notes is primarily to convince a few people, to strengthen their certainty of the fact of immortality, of the survival of the soul after the body changes which is called death."[12]

Eric J. Dingwall continued his studies in psychical research for decades to come. By the end of his life, however, his opinion about the merits of such investigation had changed. Writing in 1971 at the age of eighty-one, Dingwall reflected on the development of psychical research from the late nineteenth century to contemporary times. He spoke of Joseph B. Rhine's experiments in parapsychology in the 1930s and alluded to the resurgent popularity of the paranormal in the 1960s. Yet despite these developments in paranormal experimentation, they had only solidified Dingwall's resolve. He no longer wished to be associated with psychical researchers, parapsychologists, or anyone else "who actively support such superstitions as are today everywhere apparent. I cannot accept such responsibility."[13]

Dingwall condemned psychical research in his 1971 article as a failed science, whose adherents did not live up to empirical standards: "After sixty years experience and personal acquaintance with most of the leading parapsychologists ... I do not think I could name half a dozen whom I could call objective students." Instead, many revealed themselves to be "barefaced liars" who suppressed contrary evidence.[14] Such words seem difficult to understand when reflecting on Dingwall's long career in the field, yet they also demonstrate one investigator's now-frustrated desire that the "rational prevail over the magical."[15] Witnessing psychical research from its nineteenth-century inception to its late-twentieth-century incarnations, Dingwall had hoped, perhaps in vain, that such experimentation would lead to the downfall of "crude superstition." Instead, psychical research had, he claimed, restored occultism "from the limbo into which it had been cast" and resituated it in scientific dress. No longer supported by "criminals in red hats or ignorant priests driving devils out of sick people," the sense of the magical was buttressed by "ladies and gentlemen in white coats performing what they called experiments in what they called their laboratories."[16]

Dingwall's statements stand in direct contrast to the optimism of his younger self in the 1920s and 1930s, and they undoubtedly reflect years of frustration in attempting to forge a psychical science. Paranormal investigation remained at the margins of scientific practice, its Latourian "knot" never firmly tied. Yet, by failing to produce a narrative that demonstrated progress from superstition to rationalism, psychical research exemplified, perhaps too much so for Dingwall and others, the messiness by which knowledge is created.

Psychical researchers very purposely adopted the principles of twentieth-century science, medicine, and psychology. At the same time, circulating ideas about gender, class, and race made their way into the seance in ways less evident to investigators. The collision between these forces created the conditions of the interwar seance room. The psychical laboratory, however, was much more than a pale reflection of interwar society. It was an arena in which investigators, mediums, and spirits grappled with the norms of their surrounding society in ways that could have not occurred elsewhere. Embodied performances remained key to such struggles, providing the means by which researchers, mediums, and spirits accessed, engaged with, and disrupted empirical authority. Despite at times buttressing predominant conceptions of who could make scientific claims, these bodily enactments also created a space for these same gendered, class, and racial norms to be utterly unravelled. Unlike Donna Haraway's "modest witness," who "inhabited the culture of no culture," investigators, mediums, and spirits did not act in isolation.[17] The individuals who orchestrated paranormal investigations were by no means deluded eccentrics with ideas all their own. Rather, they operated within systems that they knew well and had embraced in order to create a science of the seance. Interwar psychical research drew into question the tenuous boundaries between science and pseudoscience, materialism and metaphysics. Providing a lens onto how such distinctions are created in the first place, paranormal investigations of this era demonstrated how science and culture could collide in ways entirely unexpected.

Notes

Groping in the Dark

1 Arthur Conan Doyle, *The History of Spiritualism*, vol. 1 (London: Cassel and Company, 1926), 5.

2 Molly McGarry, *Ghosts of Futures Past: Spiritualism and the Cultural Politics of Nineteenth-Century America* (Berkeley: University of Californian Press, 2008); Janet Oppenheim, *The Other World: Spiritualism and Psychical Research in Victorian Britain, 1850-1914* (Cambridge, UK: Cambridge University Press, 1985); Alex Owen, *The Darkened Room: Women, Power and Spiritualism in Late Victorian England* (Chicago: University of Chicago Press, 1989); Marlene Tromp, *Altered States: Sex, Nation, Drugs, and Self-Transformation in Victorian Spiritualism* (Albany: State University of New York Press, 2006); Pamela Thurschwell, *Literature, Technology and Magical Thinking, 1880-1920* (Cambridge, UK: Cambridge University Press, 2001); Tatiana Kontou and Sarah Willburn, eds., *The Ashgate Research Companion to Nineteenth-Century Spiritualism and the Occult* (Surrey, UK: Ashgate, 2012).

3 Ann Curthoys and Marilyn Lake, "Introduction," in *Connected Worlds: History in Transnational Perspective,* ed. Ann Curthoys and Marilyn Lake (Canberra: Australian National University E Press, 2005), 6–7.

4 One of Hamilton's most significant contributions to the ASPR's publications was the article he wrote about his experiences with the medium Margery. T.G. Hamilton, "Margery in Winnipeg: Three Séances of December 1926," in *Proceedings of the American Society for Psychical Research: The Margery Mediumship,* vol. 2, 556-67 (New York: American Society for Psychical Research, 1933). Hamilton also wrote articles for the British publication *Light: On Spiritualism and Psychical Research.*

5 Harry Price to Jenny O'Hara Pincock, January 16, 1931, Maines Pincock Family Fonds, GA64, series 1.1, file 4, University of Waterloo Library, Waterloo; Fidella Dismore Papa to Jenny O'Hara Pincock, May 23, 1936, Maines Pincock Family Fonds, GA64, series 1.1, file 9, University of Waterloo Library, Waterloo; Ernesto Bozzano, "William Cartheuser,

the New Medium for the Direct Voice," manuscript, trans. Fidello Dismore Papa, Maines Pincock Family Fonds, GA64, series 1.1, file 9, University of Waterloo Library, Waterloo.

6 Catharine Albanese, *A Republic of Mind and Spirit: A Cultural History of American Metaphysical Religion* (New Haven, CT: Yale University Press, 2007), 21–65; John J. Kucich, *Ghostly Communion: Cross-Cultural Spiritualism in Nineteenth-Century American Literature* (Hanover, NH: Dartmouth College Press, 2004), xi; Daphne A. Brooks, *Bodies in Dissent: Spectrality Performances of Race and Freedom, 1850-1910* (Durham, NC: Duke University Press, 2006), 15; Christopher Peterson, *Kindred Specters: Death, Mourning and American Affinity* (Minneapolis: University of Minnesota Press, 2007), 2–3; María del Pilar Blanco and Esther Peeren, "Spectropolitics: Ghosts of the Global Contemporary," in *The Spectralities Reader: Ghosts and Haunting in Contemporary Cultural Theory*, ed. María del Pilar Blanco and Esther Peeren (London: Bloomsbury, 2013), 91–92.

7 Beth A. Robertson, "Spirits of Transnationalism: Gender, Race and Cross-Correspondence in Early Twentieth-Century North America and Europe," *Gender and History* 27, 1 (2015): 151–70.

8 Stephen Kern, *The Culture of Time and Space, 1880-1918* (Cambridge, MA: Harvard University Press, 1983), 240.

9 Akira Iriye, *Global Community: The Role of International Organizations in the Making of the Contemporary World* (Berkeley: University of California Press, 2002), 9–30.

10 David Gorman, *The Emergence of International Society in the Late 1920s* (Cambridge, UK: Cambridge University Press, 2012), 3; Ann Douglas, *Terrible Honesty: Mongrel Manhattan in the 1920s* (New York: Noonday, 1995), 303.

11 Jenny Hazelgrove, *Spiritualism and British Society between the Wars* (Manchester: Manchester University Press, 2000), 204. Another study of interwar psychical research that misses this important transnational element is R. Laurence Moore, *In Search of White Crows: Spiritualism, Parapsychology and American Culture* (New York: Oxford University Press, 1977). Stan McMullin, *Anatomy of a Séance: A History of Spirit Communication in Central Canada* (Montreal and Kingston: McGill-Queen's University Press, 2004), focuses on Canadian sources but still ignores transnational collaborations, with the effect of representing individuals such as Hamilton and O'Hara Pincock as isolated and eccentric, while not taking their claims of scientific objectivity seriously.

12 The experiments of German psychical researcher Albert von Schrenck-Notzing with the medium Eva C., for instance, were questioned from at least the 1920s onward and finally exposed as fraudulent in the 1950s. Sofie Lachapelle, *Investigating the Supernatural: From Spiritism and Occultism to Psychical Research and Metaphysics in France, 1853-1931* (Baltimore, MD: John Hopkins University Press, 2011), 144–45.

13 Patricia Clavin, *Securing the World Economy: The Reinvention of the League of Nations, 1920-1946* (Oxford: Oxford University Press, 2013); Curthoys and Lake, eds., *Connected Worlds*; Daniel Laqua, *Internationalism Reconfigured: Transnational Ideas and Movements between the World Wars* (New York: I.B. Tauris, 2011); Marie Sandell, "Regional versus International: Women's Activism and Organisational Spaces in the Inter-war Period," *International History Review* 33, 4 (2011): 607–25; Waqar H. Zaidi, "'Aviation Will Either Destroy or Save Our Civilization': Proposals for the International Control of Aviation, 1920–1945," *Journal of Contemporary History* 46, 1 (2011): 150–78.

14 William James's definition of the term "radical empiricism" is found in his collection of essays *The Will to Believe and Other Essays in Popular Philosophy* (New York: Green, 1896), vii-viii, which continued to resound within a variety of religious communities well into

the twentieth century, as is outlined by Hunter Brown, *William James on Radical Empiricism and Religion* (Toronto: University of Toronto Press, 2000), 4–5.

15 Steven Shapin, "History of Science and Its Sociological Reconstructions," *History of Science* 20, 3 (1982): 196.

16 Roger Luckhurst, *The Invention of Telepathy, 1870-1901* (Oxford: Oxford University Press, 2002), 9–11; Sherrie Lynn Lyons, *Species, Serpents, Spirits and Skulls: Science at the Margins in the Victorian Age* (Albany: State University of New York Press, 2009), xi; David Patrick Thurs, *Science Talk: Changing Notions of Science in American Popular Culture* (New Brunswick, NJ: Rutgers University Press, 2007), 1.

17 Thurs, *Science Talk*, 19–20; Lyons, *Species, Serpents*, 3.

18 Ann Braude, *Radical Spirits: Spiritualism and Women's Rights in Nineteenth-Century America*, 2nd ed. (Bloomington: Indiana University Press, 2001), 23–24.

19 James H. Hyslop, "Spiritualism, Ignorance and Respectability," *Journal of the American Society for Psychical Research* 12, 12 (1918): 711.

20 Ibid., 709.

21 M.H. Halton, "Fake-Proof Apparatus to Record Spirit Voices," *Toronto Star*, 21 September 1937, 5.

22 Mark Erickson, *Science, Culture and Society: Understanding Science in the Twenty First Century* (Cambridge, UK: Polity, 2005), 3.

23 R.G.A. Dolby, "Reflections on Deviant Science," in *On the Margins of Science: The Social Construction of Rejected Knowledge*, ed. Roy Wallis (Keele, UK: University of Keele, 1979), 9–11.

24 H.M. Collins and T.J. Pinch speak to these criticisms in "The Construction of the Paranormal: Nothing Unscientific Is Happening," in *On the Margins of Science: The Social Construction of Rejected Knowledge*, ed. Roy Wallis (Keele, UK: University of Keele, 1979), 241–43.

25 Bruno Latour, *Pandora's Hope: Essays in the Reality of Science Studies* (Cambridge, MA: Harvard University Press, 1999), 100.

26 McGarry, *Ghosts of Futures Past*, 198–99.

27 Peter J. Bowler, *Science for All: The Popularization of Science in Early Twentieth-Century Britain* (Chicago: University of Chicago Press, 2009), 21, 34.

28 Rene Kollar, *Searching for Raymond: Anglicanism, Spiritualism and Bereavement between the Two World Wars* (New York: Lexington Books, 2000), x.

29 J. Malcolm Bird, *Margery the Medium* (Boston: Small, Maynard and Company, 1925), 8.

30 T.G. Hamilton, *Intention and Survival: Psychical Research Studies and the Bearing of Intentional Actions by Trance Personalities on the Problem of Human Survival*, ed. J.D. Hamilton (Toronto: Macmillan, 1942), 23.

31 Donna Haraway, "Situated Knowledges: The Science Question in Feminism and the Privilege of Partial Perspective," in *The Gender of Science*, ed. Janet A. Kourany (Upper Saddle River, NJ: Prentice Hall, 2002), 368.

32 T.G. Hamilton to Mrs. J. Bonnizer, January 14, 1931, Hamilton Family Fonds, MSS14 (A.79-41), box 5, folder 1, University of Manitoba Archives and Special Collections, Winnipeg; J.B. Rhine to J. Malcolm Bird, July 16, 1926, Margery Fonds, box 114, file 14, American Society for Psychical Research Library and Archives, New York; "Sitting at House of Dr. R.G. Crandon, 10 Lime Street," July 1, 1926, Margery Fonds, box 114, file 14, American Society for Psychical Research Library and Archives, New York.

33 Bruno Latour, *Reassembling the Social: An Introduction to Actor-Network Theory* (2005; reprint, Oxford: Oxford University Press, 2007), 24.

34 Ibid., 34.
35 Ibid., 46.
36 Owen, *Darkened Room,* 6–12.
37 Michel Foucault, "Nietzsche, Genealogy, History," in *Aesthetics, Method and Epistemology,* ed. James D. Faubion (New York: New Press, 1998), 375.
38 Saba Mahmood, "Agency, Performativity and the Feminist Subject," in *Pieties and Gender,* ed. Lene Sjorup and Hilda Romer Christensen (Leiden: Brill, 2009), 18.
39 Hamilton, *Intention and Survival,* 32.
40 This motivation on the part of the medium is suggested by Catherine Walker Bynum in her analyses of medieval female mystics and their relationship to the flesh. Caroline Walker Bynum, *Fragmentation and Redemption: Essays on Gender and the Human Body in Medieval Religion* (New York: Zone Books, 1991), 182.
41 Owen, *Darkened Room,* 10–11; Braude, *Radical Spirits,* 83.
42 Hazelgrove, *Spiritualism and British Society,* 258.
43 Hereward Carrington, *The Invisible World* (New York: Beechhurst Press and B. Ackerman, 1946), 90.
44 Ibid.
45 Evelyn Fox Keller, *Reflections on Gender and Science* (New Haven, CT: Yale University Press, 1985), 44.
46 Londa L. Schiebinger, *The Mind Has No Sex? Women in the Origins of Modern Science* (Cambridge, MA: Harvard University Press, 1989), 137.
47 Ibid., 191.
48 Wendy Mitchinson, *The Nature of Their Bodies: Women and Their Doctors in Victorian Canada* (Toronto: University of Toronto Press, 1991), 9, 49–52; Hazelgrove, *Spiritualism and British Society,* 82, 125, 153.
49 Marina Benjamin, "Introduction," in *Science and Sensibility: Gender and Scientific Enquiry, 1780-1945,* ed. Marina Benjamin (Oxford: Blackwell, 1991), 5. The Vienna Circle was a group of philosophers, scientists, and mathematicians that formed in Vienna in 1923 under the leadership of physicist Moritz Schlick. They examined scientific language and methodology and crafted a philosophy referred to as "logical positivism" or "logical empiricism." The Vienna Circle endeavoured, among other things, to perfect the concept of positivism by making it more precise and based chiefly on observation rather than on "preconceived views." Joergen Joergenson, *The Development of Logical Empiricism,* vol. 2 (Chicago: University of Chicago Press, 1951), 1–4.
50 Donna Haraway, *Modest_Witness@Second_Millennium.FemaleMan©_Meets_OncoMouse™: Feminism and Technoscience* (New York: Routledge, 1997), 32.
51 Sarah S. Richardson, *Sex Itself: The Search for Male and Female in the Human Genome* (Chicago: University of Chicago Press, 2013), 41–59, 62–76.
52 Angus McLaren, *Reproduction by Design: Sex, Robots, Trees and Test-Tube Babies in Interwar Britain* (Chicago: University of Chicago Press, 2012), 5.
53 Anne Fausto-Sterling, *Sexing the Body: Gender Politics and the Construction of Sexuality* (New York: Basic Books, 2000), 170–94; Chandak Sengoopta, *The Most Secret Quintessence of Life: Sex, Glands and Hormones, 1850-1950* (Chicago: University of Chicago Press, 2006) 69–204.
54 Alys Eve Weinbaum et al., "The Modern Girl as Heuristic Device: Collaboration, Connecitve Comparison, and Multidirectional Citation," in *The Modern Girl around the World: Consumption, Modernity, and Globalization,* ed. Alys Eve Weinbaum et al. (Durham, NC: Duke University Press, 2008), 1–2.

55 For an examination of shifting ideals of masculinity in the late nineteenth and early twentieth centuries, see Gail Bederman, *Manliness and Civilization: A Cultural History of Gender and Race in the United States, 1880-1917* (Chicago: University of Chicago Press, 1995).

56 George L. Mosse, *Fallen Soldiers: Reshaping the Memory of the World Wars* (New York: Oxford University Press, 1990), 60.

57 Angus McLaren, *The Trials of Masculinity: Policing Sexual Boundaries, 1870-1930* (Chicago: University of Chicago Press, 1997), 233–36; Joanna Bourke, *Dismembering the Male: Men's Bodies, Britain and the Great War* (Chicago: University of Chicago Press, 1996), 16–21, 30; Ben Shepherd, *A War of Nerves: Soldiers and Psychiatrists in the Twentieth Century* (Cambridge, MA: Harvard University Press, 2001), 143.

58 Bird, *Margery the Medium,* 106–8; J. Malcolm Bird, "Report to the Research Committee on Sittings with Margery Held from June 18th to October 25th 1926 with Recommendations for Future Procedure," Margery Fonds, box 119, file 2, American Society for Psychical Research Library and Archives, New York; Hamilton, *Intention and Survival,* 33–38.

59 Larry D. McCann and Peter J. Smith, "Canada Becomes Urban: Cities and Urbanization in Historical Perspective," in *Canadian Cities in Transition,* ed. Trudi Bunting and Pierre Filion (Oxford: Oxford University Press, 1991), 77–78.

60 Owen, *Darkened Room,* 8–9, 50–51. Barbara Goldsmith views this tendency of mediums as the initial sign of the weakening of spiritualism in the late nineteenth century. Barbara Goldsmith, *Other Powers: The Age of Suffrage, Spiritualism and the Scandalous Victoria Woodhull* (New York: Alfred A. Knopf, 1998), 475.

61 Henry Clay McComas, *Ghosts I Have Talked With* (Baltimore, MD: Williams and Wilkins Company, 1935), 30.

62 Seance notes, June 30, 1930, Maines Pincock Family Fonds, GA64, series 4, file 99, University of Waterloo Library, Waterloo.

63 Ryan S. Trimm, "The Times of Whiteness; Or, Race between the Postmodern and the Postcolonial," in *Postcolonial Whiteness: A Critical Reader on Race and Empire,* ed. Alfred J. Lopez (New York: State University of New York Press, 2005), 231, 247.

64 Steven Connor, "The Machine in the Ghost: Spiritualism, Technology and the Direct Voice," in *Ghosts: Deconstruction, Psychoanalysis, History,* ed. Peter Buse and Andrew Stott (London: Macmillan, 1999), 206.

65 Walter Franklin Prince, "Some Highlights," *Boston Society for Psychic Research Bulletin: The "Walter"-"Kerwin" Thumb Prints,* 22 (April 1934): 74.

66 Moore, *In Search of White Crows,* 179; Ruth Brandon, *The Spiritualists: The Passion for the Occult in the Nineteenth and Twentieth Centuries* (Buffalo: Prometheus Books, 1984), 165–89.

67 John F. Kasson, *Houdini, Tarzan and the Perfect Man: The White Male Body and the Challenge of Modernity* (New York: Hill and Wang, 2001), 145–53.

68 Ann Laura Stoler, *Along the Archival Grain: Epistemic Anxieties and Colonial Common Sense* (Princeton, NJ: Princeton University Press, 2009), 53; Antoinette Burton, *Dwelling in the Archive: Women Writing House, Home and History in Late Colonial India* (Oxford: Oxford University Press, 2003), 3–31, 138–44; Durba Ghosh, "Decoding the Nameless: Gender, Subjectivity and Historical Methodologies in Reading the Archive of Colonial India," in *A New Imperial History,* ed. Kathleen Wilson, 297–316 (Cambridge, UK: Cambridge University Press, 2004); Anjali Arondekar, "Without a Trace: Sexuality and the Colonial Archive," *Journal of the History of Sexuality* 14, 1–2 (2005): 10–27.

69 Lisa Helps, "Body, Power, Desire: Mapping Canadian Body History," *Journal of Canadian Studies* 41, 1 (2007): 130.
70 Steven Shapin and Christopher Lawrence, "Introduction: The Body of Knowledge," in *Science Incarnate: Historical Embodiments of Natural Knowledge,* ed. Steven Shapin and Christopher Lawrence (Chicago: University of Chicago Press, 1998), 10.
71 Foucault, "Nietzsche, Genealogy, History," 375–76; Michel Foucault, *Discipline and Punish: The Birth of the Prison,* trans. Alan Sheridan (New York: Vintage Books, 1977), 135–36.
72 Judith Butler, *Bodies That Matter: On The Discursive Limits of "Sex"* (New York and London: Routledge, 1993), 2.
73 James Opp, *The Lord for the Body: Religion, Medicine and Protestant Faith Healing in Canada, 1880-1930* (Montreal and Kingston: McGill-Queen's University Press, 2005), 8.
74 Lucien Febvre, "Comment reconstituer la vie affective d'autrefois? La sensibilité et l'histoire," *Annales d'histoire sociale* 3, 1-2 (1941): 5–8.
75 Robert Jütte, *A History of the Senses: From Antiquity to Cyberspace* (Cambridge, UK: Polity, 2005), 1.
76 David Howes, "Introduction," in *The Empire of the Senses: The Sensual Culture Reader,* ed. David Howes (2005; reprint, Oxford: Berg, 2006), 5; Mark M. Smith, *Sensing the Past: Seeing, Hearing, Smelling, Tasting and Touching in History* (Berkeley: University of California Press, 2007), 6.
77 Smith, *Sensing the Past,* 5.

Chapter 1: The "Scientific Self"

1 James H. Hyslop, "Signs of the Times," *Journal of the American Society for Psychical Research* 13, 2 (1919): 59.
2 Andrew Pickering, *The Mangle of Practice: Time, Agency, and Science* (Chicago: University of Chicago Press, 1995), 7, emphasis in original.
3 Steven Shapin, *A Social History of Truth: Civility and Science in Seventeenth-Century England* (Chicago: University of Chicago Press, 1994), 360.
4 Lorraine Daston and Peter Galison, *Objectivity* (New York: Zone Books, 2007), 191, 198–205.
5 W.R. Wood, "Science and Survival: A Canadian Investigator's Examination of the Evidence," *Light: A Journal of Psychic, Occult and Mystical Research* 56, 2921 (1936): 833–34.
6 Janet Oppenheim, *The Other World: Spiritualism and Psychical Research in England, 1850-1914* (Cambridge, UK: Cambridge University Press, 1985), 159.
7 Wood, "Science and Survival," 833–34.
8 Shapin, *Social History of Truth,* 65.
9 Daston and Galison, *Objectivity,* 17.
10 Ibid., 39.
11 Donna Haraway, *Modest_Witness@Second_Millennium.FemaleMan©_Meets_OncoMouse™: Feminism and Technoscience* (New York: Routledge, 1997), 23.
12 Ibid., 31.
13 W.R. Wood to Jenny O'Hara Pincock, July 8, 1936, Maines Pincock Family Fonds, GA64, file 9, series 1.1, University of Waterloo Library, Waterloo.
14 "Dr. Glen Hamilton, Psychic Researcher," *Winnipeg Tribune,* April 16, 1935, Hamilton Family Fonds, MSS14 (A.79-41), box 1, folder 4, University of Manitoba Archives and Special Collections, Winnipeg; "Psychic Field Loses Leader: Dr. T. Glen Hamilton,

Noted Investigator and Former M.L.A., Victim of Heart Ailment," *St. Catharines Standard,* undated newspaper clipping, Maines Pincock Family Fonds, GA64, series 7, file 169, University of Waterloo Library, Waterloo.

15 H.A.V. Green, "A Remarkable Materialization" (1936), Hamilton Family Fonds, MSS14 (A.79-41), box 2, folder 5, University of Manitoba Archives and Special Collections, Winnipeg; T.G. Hamilton, "A Summary of Ten Years of Psychical Research" (1931), Hamilton Family Fonds, MSS14 (A.79-41), box 2, folder 3, University of Manitoba Archives and Special Collections, Winnipeg.

16 Esyllt W. Jones, *Influenza 1918: Disease, Death and Struggle in Winnipeg* (Toronto: University of Toronto Press, 2007), 5.

17 Esyllt W. Jones, "Spectral Influenza: Winnipeg's Hamilton Family, Interwar Spiritualism and Pandemic Disease," in *Epidemic Encounters: Influenza, Society, Culture, 1918-1920,* ed. Magda Fahrni and Esyllt W. Jones (Vancouver: UBC Press, 2012), 193–95, 205–10.

18 Hamilton, "Summary of Ten Years"; T.G. Hamilton, *Intention and Survival: Psychical Research Studies and the Bearing of Intentional Actions by Trance Personalities on the Problem of Human Survival,* ed. J.D. Hamilton (Toronto: Macmillan, 1942), 18.

19 T.G. Hamilton to Dr. Mees, October 17, 1931, Hamilton Family Fonds, MSS14 (A.79-41), box 5, folder 1, University of Manitoba Archives and Special Collections, Winnipeg.

20 T.G. Hamilton to Walter Franklin Prince, January 24, 1924, Hamilton Family Fonds, MSS14 (A.79-41), box 4, folder 5, University of Manitoba Archives and Special Collections, Winnipeg.

21 T.G. Hamilton to Mrs. J. Bonnizer, January 14, 1931, Hamilton Family Fonds, MSS14 (A.79-41), box 5, folder 1, University of Manitoba Archives and Special Collections, Winnipeg.

22 Callum Brown, *Religion and Society in Twentieth-Century Britain* (Harlow, UK: Pearson Education, 2006), 75; William Westfall, *Two Worlds: The Protestant Culture of Nineteenth-Century Ontario* (Montreal and Kingston: McGill-Queen's University Press, 1989), 34; Ramsay Cook, *The Regenerators: Social Criticism in Late Victorian English Canada* (Toronto: University of Toronto Press, 1985), 25.

23 David B. Marshall, *Secularizing the Faith: Canadian Protestant Clergy and the Crisis of Belief, 1850-1940* (Toronto: University of Toronto Press, 1992), 226; Ian M. Randall, *Evangelical Experiences: A Study in the Spirituality of English Evangelicalism, 1918-1939* (Carlisle, UK: Paternoster, 1999), 69; Kevin Kee, *Revivalists: Marketing the Gospel in English Canada, 1884-1957* (Montreal and Kingston: McGill-Queen's University Press, 2006), 105–20.

24 Kee, *Revivalists,* 10; Brown, *Religion and Society,* 75; Randall, *Evangelical Experiences,* 239. The Oxford Group was a Canadian-based evangelical effort initiated by Frank Buchman in the early 1930s. For more on this group, see Kee, *Revivalists,* 96–142.

25 T.G. Hamilton to Mrs. J.A. Fisher, April 7, 1933, Hamilton Family Fonds, MSS14 (A.79-41), box 5, folder 1, University of Manitoba Archives and Special Collections, Winnipeg.

26 Ibid.

27 H.A.V. Green, "Foreword," in Hamilton, *Intention and Survival,* ix.

28 T.G. Hamilton, "The Mysteries of Teleplasm," Hamilton Family Fonds, MSS14 (A.79-41), box 15, folder 6, University of Manitoba Archives and Special Collections, Winnipeg.

29 Ann Taves, *Fits, Trances and Visions: Experiencing Religion and Explaining Experience from Wesley to James* (Princeton, NJ: Princeton University Press, 1999), 248–49; F.X. Charet, *Spiritualism and the Foundations of C.G. Jung's Psychology* (New York: State University of New York Press, 1993), 176.

30 T.G. Hamilton to L.R.G. Crandon, October 2, 1926, Hamilton Family Fonds MSS14 (A.79-41), box 4, folder 4, University of Manitoba Archives and Special Collections, Winnipeg.

31 T.G. Hamilton, "Teleplasmic Phenomenon in Winnipeg," *Psychic Science: Quarterly Transactions of the British College* 8, 3 (1929): 180.

32 John Lardas Modern, *Secularism in Antebellum America* (Chicago: University of Chicago Press, 2011), 7–9, 16–19.

33 Hamilton, "Teleplasmic Phenomenon in Winnipeg," 179.

34 Ibid., 180.

35 Ibid., 184.

36 T.G. Hamilton, "Telekinesis" (1928), Hamilton Family Fonds, MSS14 (A.79-41), box 9, folder 13, University of Manitoba Archives and Special Collections, Winnipeg.

37 Ibid.

38 Ibid.

39 Hamilton, "Teleplasmic Phenomenon in Winnipeg," 184–85.

40 J. Malcolm Bird, *Margery the Medium* (Boston: Small, Maynard and Company, 1926), 13.

41 Ibid., 16.

42 Ibid. 16–20, 50–51, 55.

43 Other members included psychologist Gardner Murphy and his assistant Harry Helson, a doctoral candidate in psychology who was also at Harvard and joined the group in November.

44 Bird, *Margery the Medium*, 36, 51, 101, 103, 134–38.

45 E.J. Dingwall, *How to Go to a Medium: A Manual of Instruction* (London: Kegan Paul, Trench, Trubner and Company, 1922), 6.

46 E.J. Dingwall, "A Report of a Series of Sittings with the Medium Margery, June 1926," in *Proceedings of the Society for Psychical Research*, vol. 36 (Glasgow, UK: Robert MacLehose and Company, 1928), 80.

47 Ibid., 149–52.

48 John F. Kasson, *Houdini, Tarzan and the Perfect Man: The White Male Body and the Challenge of Modernity* (New York: Hill and Wang, 2001), 143–53; Marina Warner, *Phantasmagoria: Spirit Visions, Metaphors, and Media into the Twenty-First Century* (Oxford: Oxford University Press, 2006), 230.

49 Walter Franklin Prince, "Some Highlights," *Boston Society for Psychic Research Bulletin: "Walter"-"Kerwin" Thumb Prints*, 22 (April 1934), 74; "Famed Medium, Ontario Born 'Margery,' Dies," *St. Catharines Standard*, November 3, 1941, Maines Pincock Family Fonds, GA64, file 169, series 1.1, University of Waterloo Library, Waterloo.

50 Henry Clay McComas, *Ghosts I Have Talked With* (Baltimore, MD: Williams and Wilkins Company, 1935), 110.

51 One of the debates in which Crandon engaged within the *Journal of the American Society for Psychical Research* was with French researcher René Sudre, documented in Le Roi Crandon, "The Margery Mediumship: Supernormal Cognition, Subjective and Objective," *Journal of the American Society for Psychical Research* 20, 6 (1926): 321–33; and René Sudre, "Margery and Spiritism," *Journal of the American Society for Psychical Research* 20, 9 (1926): 524–31.

52 Crandon, "Margery Mediumship," 324.

53 Dingwall, "Report of a Series," 84.

54 T.G. Hamilton to L.R.G. Crandon, May 5, 1925, Hamilton Family Fonds, MSS14 (A.79-41), box 4, file 4, University of Manitoba Archives and Special Collections, Winnipeg.

55 T.G. Hamilton, "Margery in Winnipeg: Three Séances of December 1926," in *Proceedings of the American Society for Psychical Research: The Margery Mediumship*, vol. 2, 556–67 (New York: American Society for Psychical Research, 1933). These apparatuses are discussed in more detail in Chapter 5.

56 Ibid., 558.

57 T.G. Hamilton to J. Malcolm Bird, January 18, 1927, Hamilton Family Fonds, MSS14 (A.79-41), box 4, folder 5, University of Manitoba Archives and Special Collections, Winnipeg.

58 T.G. Hamilton to Mr. A.W. Neill, March 27, 1929, Hamilton Family Fonds, MSS14 (A.79-41), box 4, folder 5, University of Manitoba Archives and Special Collections, Winnipeg.

59 Dingwall, *How to Go to a Medium*, 5.

60 Bird, *Margery the Medium*, 5–6.

61 Henry Clay McComas and Mary Winona McComas, *The McComas Saga: A Family History down to the Year 1950* (Baltimore, MD: McComas, 1950), 256–58.

62 McComas, *Ghosts I Have Talked With*, 29.

63 Ibid., 29–30.

64 Ibid., 15.

65 Haraway, *Modest_Witness*, 24.

66 Ibid., 31.

67 Ruth Watts explains how the interwar years witnessed a revived interest in gender-differentiated public school curriculum, which obstructed the pursuit of scientific subjects by young girls and women. She also points out how the few women who did obtain higher education in the sciences struggled to gain viable employment in their respective fields. Ruth Watts, *Women and Science: A Social and Cultural History* (London: Routledge, 2007), 134, 146, 165–66.

68 L. Hamilton, "Steps in the Hamilton Inquiry" (c. 1945), Hamilton Family Fonds, MSS14 (A.79-41), box 15, folder 6, University of Manitoba Archives and Special Collections, Winnipeg.

69 Ibid.

70 Ibid.

71 Seance notes, July 24, 1921, quoted in T.G. Hamilton, "Second Period of Development," Hamilton Family Fonds, MSS14 (A.79-41), box 9, folder 15, University of Manitoba Archives and Special Collections, Winnipeg.

72 Ibid.

73 L. Hamilton, "Steps in the Hamilton Inquiry."

74 Ibid.

75 Hamilton, "Teleplasmic Phenomenon in Winnipeg," 180.

76 Evelyn Fox Keller, *Reflections on Gender and Science* (New Haven, CT: Yale University Press, 1985), 7.

77 T.G. Hamilton to Isabel M. Stewart, April 29, 1931, Hamilton Family Fonds, MSS14 (A.79-41), box 5, folder 1, University of Manitoba Archives and Special Collections, Winnipeg.

78 Ann Braude, *Radical Spirits: Spiritualism and Women's Rights in Nineteenth-Century America*, 2nd ed. (Bloomington: Indiana University Press, 2001), 82–83; Alex Owen, *The Darkened Room: Women, Power, and Spiritualism in Late Victorian England* (Chicago: University of Chicago Press, 1989), 7; Phyllis Mack, *Visionary Women: Ecstatic Prophecy in Seventeenth-Century England* (Berkeley: University of California Press, 1992), 24, 45;

Heather D. Curtis, *Faith in the Great Physician: Suffering and Divine Healing in American Culture* (Baltimore, MD: John Hopkins University Press, 2007), 38; Joanna Dean, *Religious Experience and the New Woman: The Life of Lily Dougall* (Bloomington: Indiana University Press, 2007), 9.

79 L. Hamilton, "Steps in the Hamilton Inquiry."

80 Ibid.

81 H.A.V. Green, "A Short Account of the Mary M. Photographs," *Light: A Journal of Psychical, Occult and Mystical Research* 49, 2521 (1929): 207.

82 L. Hamilton, "Steps in the Hamilton Inquiry," 20.

83 Lorraine Code, *What Can She Know? Feminist Theory and the Construction of Knowledge* (Ithaca, NY: Cornell University Press, 1991), 10.

84 Shapin, *Social History of Truth*, 360.

85 E.E. Dudley, "Miscellaneous Episodes of Late 1926," in *Proceedings of the American Society for Psychical Research: The Margery Mediumship*, vol. 2 (New York: American Society for Psychical Research, 1933), 545–52.

86 Seance notes, September 25, 1927, Maines Pincock Family Fonds, GA64, series 4, file 79, University of Waterloo Library, Waterloo; Jenny O'Hara Pincock to Minnie Maines, September 26, 1927, Maines Pincock Family Fonds, GA64, series 1, file 2, University of Waterloo Library, Waterloo.

87 Pincock to Maines, September 26, 1927.

88 R.H. Saunders to Jenny O'Hara Pincock, July 7, 1928, Maines Pincock Family Fonds, GA64, series 1, file 2, University of Waterloo Library, Waterloo; B.F. Austin, "Foreword," in Jenny O'Hara Pincock, *Trails of Truth* (Los Angeles, CA: Austin, 1930), 4.

89 Jenny O'Hara Pincock, *Trails of Truth* (Los Angeles, CA: Austin, 1930), 27.

90 Ibid., 7, 11–24, 27.

91 Ibid., 13, 15.

92 Ibid., 24.

93 W.E. Ross to Jenny O'Hara Pincock, December 12, 1930, Maines Pincock Family Fonds, GA64, series 1, file 3, University of Waterloo Library, Waterloo.

94 Harry Price to Jenny O'Hara Pincock, January 16, 1931, Maines Pincock Family Fonds, GA64, series 1, file 4 University of Waterloo Library, Waterloo.

95 Fidella Dismore Papa to Jenny O'Hara Pincock, May 23, 1936, Maines Pincock Family Fonds, GA64, series 1, file 9, University of Waterloo Library, Waterloo.

96 Ernesto Bozzano, "William Cartheuser, the New Medium for the Direct Voice," manuscript, trans. Fidello Dismore Papa, Maines Pincock Family Fonds, GA64, series 1, file 9, University of Waterloo Library, Waterloo.

97 O'Hara Pincock, *Trails of Truth*, 11–19.

98 Ibid., 7.

99 Alison Winter, "A Calculus of Suffering: Ada Lovelace and the Bodily Constraints on Women's Knowledge in Early Victorian England," in *Science Incarnate: Historical Embodiments of Natural Knowledge*, ed. Steven Shapin and Christopher Lawrence (Chicago: University of Chicago Press, 1998), 207.

100 Seance notes, August 5, 1928, Hamilton Family Fonds, MSS14 (A.79-41), box 15, folder 12, University of Manitoba Archives and Special Collections, Winnipeg.

101 Mary Jacobus, Evelyn Fox Keller, and Sally Shuttleworth, "Introduction," in *Body Politics: Women and the Discourses of Science*, ed. Mary Jacobus, Evelyn Fox Keller, and Sally Shuttleworth (New York: Routledge, 1990), 7.

102 Hereward Carrington, *The Invisible World* (New York: Beechhurst Press and B. Ackerman, 1946), 7.
103 Bruno Latour and Steven Woolgar, *Laboratory Life: The Construction of Scientific Facts* (London: Sage, 1979), 238.
104 Haraway, *Modest_Witness,* 5.
105 Lily Dale assembly pamphlet, season 1931, Maines Pincock Family Fonds, GA64, series 5, file 158, University of Waterloo Library, Waterloo.
106 Patricia Jansen, *Wild Things: Nature, Culture and Tourism in Ontario, 1790-1914* (Toronto: University of Toronto Press, 1995), 112–13.
107 Louis S. Vosburgh Library, undated Lily Dale advertisement pamphlet, Maines Pincock Family Fonds, GA64, series 5, file 158, University of Waterloo Library, Waterloo.
108 McComas, *Ghosts I Have Talked With,* 89–90.
109 Amelia Holden to Jenny O'Hara Pincock, September 20, 1935, Maines Pincock Family Fonds, GA64, series 1, file 8, University of Waterloo Library, Waterloo.
110 McComas, *Ghosts I Have Talked With,* 31.
111 Ibid., 33.
112 Ibid., 49–50.
113 Ibid., 51.
114 Bird, *Margery the Medium,* 34, 92–102, 115.
115 Ibid., 95, 104.
116 Ibid., 107–8.
117 Ibid., 110–15.
118 Crosbie Smith and John Agar, "Introduction: Making Space for Science," in *Making Space for Science: Territorial Themes in the Shaping of Knowledge,* ed. Crosbie Smith and John Agar (Houndmills, UK: Macmillan, 1998), 9–12.
119 David Gooding, "History in the Laboratory: Can We Tell What Really Went On?" in *The Development of the Laboratory: Essays on the Place of Experiment in Industrial Civilization,* ed. Frank A.J.L. James (Houndmills, UK: Macmillan, 1989), 63.
120 Carrington, *Invisible World,* 7.
121 Braude, *Radical Spirits,* 24.
122 Henri Lefebvre, *The Production of Space,* trans. Donald Nicholson-Smith (1974; reprint, Oxford: Blackwell, 2000), 26.
123 Dingwall, "Report of a Series," 86–88.
124 Cynthia Comacchio, *The Infinite Bonds of Family: Domesticity in Canada, 1850-1940* (Toronto: University of Toronto Press, 1999), 20–24; Peter Ward, *A History of Domestic Space: Privacy and the Canadian Home* (Vancouver: UBC Press, 1999), 62.
125 Ward, *History of Domestic Space,* 60–71.
126 "First Experiment," March 26, 1923, Hamilton Family Fonds, PC12 (A.79-41), box 9, folder 1, "Telekinesis," University of Manitoba Archives and Special Collections, Winnipeg.
127 T.G. Hamilton, "Levitations by Psychic Force (Telekinesis)," Hamilton Family Fonds, PC12 (A.79-41), box 8, folder 1, "Telekinesis," University of Manitoba Archives and Special Collections, Winnipeg.
128 Green, "Foreword," x.
129 T.G. Hamilton, "On Trance" (1927?), Hamilton Family Fonds, MSS14 (A.79-41), box 9, folder 11, Hamilton Family Fonds, Winnipeg.
130 Haraway, *Modest_Witness,* 24.
131 Daston and Galison, *Objectivity,* 191.

Chapter 2: Otherwordly Subjects

1 Waldemar Kaempffert, "The Scientist," *Journal of the American Society for Psychical Research* 14, 10 (1920): 485.
2 Donna J. Haraway, *Modest_Witness@Second_Millennium.FemaleMan©_Meets_OncoMouse™: Feminism and Technoscience* (New York: Routledge, 1997), 25.
3 Alex Owen, *The Darkened Room: Women, Power, and Spiritualism in Late Nineteenth Century England* (Chicago: University of Chicago Press, 1989), xii, 10–11; Ann Braude, *Radical Spirits: Spiritualism and Women's Rights in Nineteenth-Century Century America,* 1st ed. (Boston: Beacon, 1989), 83–85; Diana Basham, *The Trial of Woman: Feminism and the Occult Sciences in Victorian Literature and Society* (Basingstoke, UK: Macmillan, 1992), 130–31.
4 In this regard, mediums were similar, for instance, to mesmerist subjects, according to Alison Winter, *Mesmerized: Powers of Mind in Victorian Britain* (Chicago: University of Chicago Press, 1998), 62–63.
5 Molly McGarry, *Ghosts of Futures Past: Spiritualism and the Cultural Practices of Nineteenth-Century America* (Berkeley: University of California Press, 2008), 154; Marlene Tromp, *Altered States: Sex, Nation, Drugs, and Self-Transformation in Victorian Spiritualism* (Albany: State University of New York Press, 2006), 27.
6 Tromp, *Altered States,* 1.
7 Owen, *Darkened Room,* 6–8.
8 Charles Richet, *Thirty Years of Psychic Research: A Treatise on Metaphysics,* trans. Stanley Debrath (Toronto: Macmillan, 1923), 64.
9 T.G. Hamilton, "Variations of Trance Condition," Hamilton Family Fonds, MSS14 (A.79-41), box 9, folder 11, University of Manitoba Archives and Special Collections, Winnipeg.
10 Ibid.
11 T.G. Hamilton, "Article 3: Teleplasm Found in Psychic Study" (c. 1930), Hamilton Family Fonds MSS14 (A.79-41), box 15, folder 6, University of Manitoba Archives and Special Collections, Winnipeg.
12 T.G. Hamilton, "A Study of the Winnipeg Group Mediumship in Its Relation to the Dawn Teleplasms," *Psychic Research* 29, 2 (1934): 118.
13 E.E. Dudley, "Telekinesis in 1928 and 1929," in *Proceedings of the American Society for Psychical Research: The Margery Mediumship,* vol. 2 (New York: American Society for Psychical Research, 1933), 679.
14 Melanie King, *The Dying Game: A Curious History of Death* (Oxford: Oneworld, 2008), 17–18.
15 Lillian Hamilton, "Report of Sitting," February 10, 1932, Hamilton Family Fonds, MSS14 (A.79-41), box 16, folder 8, University of Manitoba Archives and Special Collections, Winnipeg.
16 "Sitting at 10 Lime Street," January 14, 1926, Margery Fonds, box 119, file 7, American Society for Psychical Research Library and Archives, New York; "Sitting at 10 Lime Street," March 18, 1926, Margery Fonds, box 119, file 7, American Society for Psychical Research Library and Archives, New York.
17 Phyllis Mack has argued that the female visionary in the seventeenth century was cast as a passive and receptive "empty vessel" so that the spirit could work through her. Phyllis Mack, *Visionary Women: Ecstatic Prophecy in Seventeenth-Century England* (Berkeley: University of California Press, 1992), 33. This same principle of femininity, religion, and

passivity was carried into the late twentieth century, as shown by scholars such as Elaine J. Lawless, *God's Peculiar People: Women's Voices and Folk Tradition in a Pentecostal Church* (Lexington: University Press of Kentucky, 1988), 77.

18 Owen, *Darkened Room,* 7–10.

19 Alys Eve Weinbaum et al., "The Modern Girl as Heuristic Device," in *The Modern Girl around the World: Consumption, Modernity, and Globalization,* ed. Alys Eve Weinbaum et al. (Durham, NC: Duke University Press, 2008), 1–2, 12.

20 Veronica Strong-Boag, *The New Day Recalled: Lives of Girls and Women in English Canada, 1919-1939* (Toronto: Copp Clark Pittman, 1988), 2; Alison Prentice et al., *Canadian Women: A History* (Toronto: Harcourt Brace Jovanovich, 1988), 240.

21 Tamara Myers, *Caught: Montreal's Modern Girls and the Law, 1869-1945* (Toronto: University of Toronto Press, 2006), 58–59.

22 Mary Louise Roberts, *Civilization without Sexes: Reconstructing Gender in Postwar France, 1917-27* (Chicago: University of Chicago Press, 1994), 5–9.

23 Ibid., 5–14; Weinbaum et al., "Modern Girl," 1–2.

24 Sarah Deutsch, *Women and the City: Gender, Space and Power in Boston, 1870-1940* (Oxford: Oxford University Press, 2002), 79–91.

25 J. Malcolm Bird, "Paraffin Gloves under the Co-mediumship of Dr. Hardwicke," in *Proceedings of the American Society for Psychical Research: The Margery Mediumship,* vol. 1 (New York: American Society for Psychical Research, 1928), 447.

26 Walter Franklin Prince, "Margery Case Notes," Margery Fonds, box 117, file 4, American Society for Psychical Research Library and Archives, New York; Henry Clay McComas, *Ghosts I Have Talked With* (Baltimore, MD: Williams and Wilkins Company, 1935), 121.

27 Langsworth Paxton to J.B. Rhine, July 31, 1926, Margery Fonds, box 114, file 14, American Society for Psychical Research Library and Archives, New York.

28 R. Laurence Moore, *In Search of White Crows: Spiritualism, Parapsychology and American Culture* (New York: Oxford University Press, 1977), 188; H.M. Collins and T.J. Pinch, "The Construction of the Paranormal: Nothing Unscientific Is Happening," in *On the Margins of Science: The Social Construction of Rejected Knowledge,* ed. Roy Wallis (Keele, UK: University of Keele, 1979), 241–43; Louisa E. Rhine, *Mind over Matter* (New York: Macmillan, 1970), 4, 70–73.

29 J.B. Rhine and Louisa E. Rhine, "One Evening's Observation on the Margery Mediumship," *Journal of Abnormal and Social Psychology* 21, 4 (1927): 417.

30 Le Roi Crandon, "The Margery Mediumship: Supernormal Cognition, Subjective and Objective," *Journal of the American Society for Psychical Research* 20, 6 (1926): 325.

31 J. Malcolm Bird, *Margery the Medium* (Boston: Small, Maynard and Company, 1925), 220.

32 McComas, *Ghosts I Have Talked With,* 141.

33 J.B. Rhine to J. Malcolm Bird, July 16, 1926, Margery Fonds, box 114, file 14, American Society for Psychical Research Library and Archives, New York.

34 McGarry, *Ghosts of Futures Past,* 154.

35 Vieda Skultans, "Mediums, Controls and Eminent Men," in *Women's Religious Experience,* ed. Pat Holden (London: Croom Helm, 1983), 23; E.J. Dingwall, *Some Human Oddities: Studies in the Queer, the Uncanny and the Fanatical* (London: Home and Van Thal, 1947), 122–23.

36 I have written about this effect elsewhere with regard to other male mediums and their experiences in the twentieth-century seance room. Beth A. Robertson, "Spirits of

Transnationalism: Gender, Race and Cross-Correspondence in Early Twentieth-Century North America and Europe," *Gender and History* 27, 1 (2015): 151–52, 157–62, 164–65.

37 McComas, *Ghosts I Have Talked With,* 32.

38 Ibid.; Hereward Carrington, *The Invisible World* (New York: Beechhurst Press and B. Ackerman, 1946), 68.

39 J. Malcolm Bird, quoted in E.E. Dudley, "Miscellaneous Episodes of Late 1926," in *Proceedings of the American Society for Psychical Research: The Margery Mediumship,* vol. 2 (New York: American Society for Psychical Research, 1933), 546.

40 McComas, *Ghosts I Have Talked With,* 31; Dudley, "Miscellaneous Episodes," 548; Carrington, *Invisible World,* 48.

41 McComas, *Ghosts I Have Talked With,* 37.

42 Evelyn Fox Keller, *Secrets of Life, Secrets of Death: Essays on Language, Gender and Science* (New York: Routledge, 1992), 41.

43 Jenny O'Hara Pincock, *Trails of Truth* (Los Angeles, CA: Austin, 1930), 30, 37; Seance notes, September 25, 1927, Maines Pincock Family Fonds, GA64, series 4, file 79, University of Waterloo Library, Waterloo. For more on this group's healing practices, see Beth A. Robertson, "Radiant Healing: Gender, Belief, and Alternative Medicine in St. Catharines, Ontario, Canada 1927-1935," *Nova Religio: The Journal of Alternative and Emergent Religions* 18,1 (2014): 16–36.

44 O'Hara Pincock, *Trails of Truth,* 30.

45 Ibid.

46 Seance notes, July 1, 1928, Maines Pincock Family Fonds, GA64, series 4, file 81, University of Waterloo Library, Waterloo; Seance notes, March 29, 1929, Maines Pincock Family Fonds, GA64, series 4, file 92, University of Waterloo Library, Waterloo.

47 Seance notes, May 6, 1931, Maines Pincock Family Fonds, GA64, series 4, file 108, University of Waterloo Library, Waterloo.

48 Seance notes, October 5, 1932, Maines Pincock Family Fonds, GA64, series 4, file 136, University of Waterloo Library, Waterloo.

49 Gail Bederman, *Manliness and Civilization: A Cultural History of Gender and Race in the United States, 1880-1917* (Chicago: University of Chicago Press, 1995), 5–17.

50 Ibid., 19.

51 Christina Simmons, *Making Marriage Modern: Women's Sexuality from the Progressive Era to World War II* (Oxford: Oxford University Press, 2009), 11–13, 142–43, 182–83.

52 Laura King, *Family Men: Fatherhood and Masculinity in Britain, 1914-1960* (Oxford: Oxford University Press, 2015), 2–4, 51–88.

53 Jessica Meyer, *Men of War: Masculinity and the First World War in Britain* (Hampshire, UK: Palgrave Macmillan, 2009), 3; Ben Shepherd, *A War of Nerves: Soldiers and Psychiatrists in the Twentieth Century* (Cambridge, MA: Harvard University Press, 2001), xix, 41; Angus McLaren, *The Trials of Masculinity: Policing Sexual Boundaries, 1870-1930* (Chicago: University of Chicago Press, 1997), 233.

54 Joanna Bourke, *Dismembering the Male: Men's Bodies, Britain and the Great War* (Chicago: University of Chicago Press, 1996), 30–33.

55 Minnie Maines to Jenny O'Hara Pincock, October 6, 1931, Maines Pincock Family Fonds, GA64, series 1.1, file 4, University of Waterloo Library, Waterloo.

56 Jenny O'Hara Pincock to Henry Remmers, April 5, 1944, Maines Pincock Family Fonds, GA64, series 1.1, file 17, University of Waterloo Library, Waterloo.

57 Ibid.

58 Margaret L. Hamilton, "Notes on Mr. Harry A.V. Green, Co-worker with Dr. T. Glen Hamilton," Hamilton Family Fonds, MSS14 (A.79-41), box 1, folder 4, University of Manitoba Archives and Special Collections, Winnipeg.
59 Hamilton, "Study of the Winnipeg Group," 126.
60 Ibid.
61 Ibid.
62 "The Pirate Speaks through Ewan," Seance notes, February 22, 1930, Hamilton Family Fonds, MSS14 (A.79-41), box 16, folder 2, University of Manitoba Archives and Special Collections, Winnipeg.
63 Margaret L. Hamilton, "Notes on Mr. Harry A.V. Green"; T.G. Hamilton, "Description and Initials of a Man Who Is to Be Photographed," September 15, 1929, Hamilton Family Fonds, MSS14 (A.79-41), box 16, folder 1, University of Manitoba Archives and Special Collections, Winnipeg.
64 Joy Dixon, *The Divine Feminine: Theosophy and Feminism in England* (Baltimore, MD: John Hopkins University Press, 2001), 84–86.
65 "The Pirate Speaks through Ewan"; Margaret L. Hamilton, "Notes on Mr. Harry A.V. Green"; Marina Warner, *Phantasmagoria: Spirit Visions, Metaphors, and Media into the Twenty-First Century* (Oxford: Oxford University Press, 2006), 227.
66 Seance notes, November 12, 1933, Hamilton Family Fonds, MSS14 (A.79-41), box 17, folder 1, University of Manitoba Archives and Special Collections, Winnipeg.
67 Hamilton, "Study of the Winnipeg Group," 118.
68 W.E. Hobbs, "Walter Preparing Something," Seance notes, January 12, 1930, Hamilton Family Fonds, MSS14 (A.79-41), box 16, folder 2, University of Manitoba Archives and Special Collections, Winnipeg.
69 Hamilton, "Study of the Winnipeg Group," 118.
70 "At Last - Materialization Appears," Seance notes, November 29, 1931, Hamilton Family Fonds, MSS14 (A.79-41), box 16, folder 7, University of Manitoba Archives and Special Collections, Winnipeg.
71 Fox Keller, *Secrets of Life,* 40–41, explains the development of a gender politics that ascribed power to scientific men over a natural, feminine other.
72 H.A.V. Green, "Foreword," in T.G. Hamilton, *Intention and Survival: Psychical Research Studies and the Bearing of Intentional Actions by Trance Personalities on the Problem of Human Survival,* ed. J.D. Hamilton (Toronto: Macmillan, 1942), ix-xiv.
73 Margaret L. Hamilton, "Notes on Mr. Harry A.V. Green."
74 John F. Kasson, *Houdini, Tarzan and the Perfect Man: The White Male Body and the Challenge of Modernity* (New York: Hill and Wang, 2001), 10.
75 Ibid., 222.
76 Owen, *Darkened Room,* 49–50.
77 Reinhold Kramer and Tom Mitchell, *When the State Trembled: How A.J. Andrews and the Citizens' Committee Broke the Winnipeg General Strike* (Toronto: University of Toronto Press, 2010), 7–9.
78 Craig Heron, "The Workers Revolt, 1917-1925," in *Labouring Canada: Class, Gender, and Race in Canadian Working Class History,* ed. Bryan D. Palmer and Joan Sangster (Oxford: Oxford University Press, 2008), 138–41; Bryan D. Palmer, *Working Class Experience: Rethinking the History of Canadian Labour, 1800-1991,* 2nd ed. (Toronto: McClelland and Stewart, 1992), 200–1; Carmela Patrias and Larry Savage, *Confrontation, Struggle and Transformation: Organized Labour in the St. Catharines Area* (Edmonton: Athabasca

University Press, 2007), 10–13; David Bright, *The Limits of Labour: Class Formation and the Labour Movement in Calgary, 1883-1929* (Vancouver: UBC Press, 1998), 146-47.

79 Bruce Levine, *Who Built America? Working People and the Nation's Economy, Politics, Culture and Society,* vol. 2 (New York: Pantheon Books, 1992), 260–61; Joanna Bourke, *Working-Class Cultures in Britain, 1890-1960: Gender, Class and Ethnicity* (London: Routledge, 1994), 179–80; Mike Savage, *The Remaking of the British Working Class, 1840-1940* (London: Routledge, 1994), 80.

80 David R. Roediger, *Towards the Abolition of Whiteness: Essays on Race, Politics and Working Class History* (New York: Verso, 1994), 21–23, 61–68.

81 Bourke, *Working-Class Cultures,* 208–11.

82 Joy Parr, *The Gender of Breadwinners: Women, Men and Change in Two Industrial Towns, 1880-1950* (Toronto: University of Toronto Press, 1990), 235, 238; Elizabeth Faue, "Re-imagining Labor: Gender and New Directions in Labour and Working-Class History," in *Re-thinking U.S. Labour History: Essays on the Working-Class Experience, 1756-2009,* ed. Donna T. Haverty-Stacke and Daniel J. Walkowitz (New York: Continuum International Publishing Group, 2010), 268; Bourke, *Working-Class Cultures,* 130–31.

83 Parr, *Gender of Breadwinners,* 66; Margaret Hobbs and Joan Sangster, *The Woman Worker, 1926-29* (St. John's, NL: Canadian Committee on Labour History, 1999), 36–39; Katrina Srigley, *Breadwinning Daughters: Young Working Women in a Depression-Era City, 1929-1939* (Toronto: University of Toronto Press, 2010), 21–22; John Benson, *The Working Class in Britain, 1850-1939* (London: I.B. Taurus, 2003), 177.

84 Fox Keller, *Secrets of Life,* 40–41.

85 T.G. Hamilton, "Attitude of E.M. and the Experimenters Towards the Trance Personalities" (c. 1926), Hamilton Family Fonds, MSS14 (A.79-41), box 9, folder 9, University of Manitoba Archives and Special Collections, Winnipeg.

86 T.G. Hamilton, "Teleplasmic Phenomenon in Winnipeg," *Psychic Science: Quarterly Transactions of the British College* 8, 3 (1929): 183; Hamilton, "Attitude of E.M."

87 Hamilton, "Attitude of E.M."; Hamilton, "Teleplasmic Phenomenon in Winnipeg," 180.

88 Eva Broad to Lillian Hamilton, January 12, 1925, Hamilton Family Fonds, MSS14 (A.79-41), box 1, folder 4, University of Manitoba Archives and Special Collections, Winnipeg.

89 "Prediction re W.O. Hamilton's Death by Mrs. Mary Marshall, December 2, 1924," Hamilton Family Fonds, MSS14 (A.79-41), box 1, folder 4, University of Manitoba Archives and Special Collections, Winnipeg.

90 Mary D. Marshall to Lillian Hamilton, July 1, 1945, Hamilton Family Fonds, MSS14 (A.79-41), box 5, folder 4, University of Manitoba Archives and Special Collections, Winnipeg.

91 Ibid., 4.

92 Ibid., 8–10.

93 "The Margery Mediumship: Dr. Crandon's Statement by Our Own Representative," *The Two Worlds,* December 22, 1933, 982.

94 Owen, *Darkened Room,* 60.

95 "Famed Ontario Born Margery Dies," *St. Catharines Standard,* November 3, 1941, Maines Pincock Family Fonds, GA64, series 6, file 169, University of Waterloo Archives and Special Collections, Waterloo; Thomas R. Tietze, *Margery* (New York: Harper and Row, 1973), 2–6; Bird, *Margery the Medium,* 13.

96 R.J. Tillyard, "Some Recent Personal Experiences with Margery," *Journal of the American Society for Psychical Research* 20, 12 (1926): 709.

97 E.J. Dingwall, "A Report of a Series of Sittings with the Medium Margery, June 1926," in *Proceedings of the Society for Psychical Research,* vol. 36 (Glasgow, UK: Robert MacLehose and Company, 1928), 98; T.G. Hamilton, "Margery in Winnipeg: Three Séances of December 1928," in *Proceedings of the American Society for Psychical Research: The Margery Mediumship,* vol. 2 (New York: American Society for Psychical Research, 1933), 560–61.
98 Bird, *Margery the Medium,* 85, 105; "Sitting at 10 Lime Street," September 15, 1926, Margery Fonds, box 119, file 7, American Society for Psychical Research Library and Archives, New York.
99 Bird, quoted in Dudley, "Miscellaneous Episodes," 545–46.
100 Bird, *Margery the Medium,* 26.
101 Seance notes, September 19, 1926, Margery Fonds, box 119, file 7, American Society for Psychical Research Library and Archives, New York. Also reported by Daniel Day Walton, "Notable Sittings in 1926: Houdini's Last Contact with the Case," in *Proceedings of the American Society for Psychical Research: The Margery Mediumship,* vol. 2 (New York: American Society for Psychical Research, 1933), 536.
102 Hamilton compiled many lengthy records of the evidence that such personalities bestowed, one of which is T.G. Hamilton, "R.L.S., D.L. and W.T.S. Transmissions by Trance-Automatisms and Trance-Visions through E.M. Verifications, July 28th 1926 to Nov. 6, 1927, Including Examples of Deep-Trance Automatic Scripts," Hamilton Family Fonds, MSS14 (A.79-41), box 15, folder 4, University of Manitoba Archives and Special Collections, Winnipeg.
103 T.G. Hamilton, "Teleplasmic Phenomenon in Winnipeg," 185.
104 "First Conversation with Walter," Seance notes, March 4, 1928, Hamilton Family Fonds, MSS14 (A.79-41), box 15, folder 11, University of Manitoba Archives and Special Collections, Winnipeg.
105 "Regular Sitting," Seance notes, April 1, 1928, Hamilton Family Fonds, MSS14 (A.79-41), box 15, folder 11, University of Manitoba Archives and Special Collections, Winnipeg.
106 "The Young Man Asks for an Electric Bell," Seance notes, March 15, 1928, Hamilton Family Fonds, MSS14 (A.79-41), box 15, folder 11, University of Manitoba Archives and Special Collections, Winnipeg.
107 Ibid.
108 Seance notes, March 22, 1928, Hamilton Family Fonds, MSS14 (A.79-41), box 15, folder 11, University of Manitoba Archives and Special Collections, Winnipeg.
109 Seance notes, April 1 1928, Hamilton Family Fonds, MSS14 (A.79-41), box 15, folder 11, University of Manitoba Archives and Special Collections, Winnipeg.
110 T.G. Hamilton to Miss L.E. Johnson, October 23, 1931, Hamilton Family Fonds. MSS14 (A.79-41), box 5, folder 1, University of Manitoba Archives and Special Collections, Winnipeg.
111 Hamilton, "Teleplasmic Phenomenon in Winnipeg," 184–85.
112 Hamilton, "Margery in Winnipeg," 558.
113 Hamilton, "Teleplasmic Phenomenon in Winnipeg," 185.
114 T.G. Hamilton, "A Review of Volume XXII of the Proceedings of the American Society for Psychical Research" (1931), Hamilton Family Fonds, MSS14 (A.79-41), box 2, file 4, University of Manitoba Archives and Special Collections, Winnipeg.
115 Hamilton, "Teleplasmic Phenomenon in Winnipeg," 185.
116 T.G. Hamilton, "The C.H. Spurgeon Case: Evidence Pointing to Survival and Continued Activity," *Light: A Journal of Psychical, Occult and Mystical Research* 53, 2752 (1933): 628.

117 Seance notes, November 27, 1932, Hamilton Family Fonds, MSS14 (A.79-41), box 16, folder 10, University of Manitoba Archives and Special Collections, Winnipeg.
118 "An Indian Origin for Modern Spiritualism," *Toronto Star,* May 22, 1928, 19.
119 Shari M. Huhndorf, *Going Native: Indians in the American Cultural Imagination* (Ithaca, NY: Cornell University Press, 2001), 14.
120 "A Children's Party Held on March 16th, 1929 at 47 Church Street St. Catharines Ontario," Maines Pincock Family Fonds, GA64, series 4, file 91, University of Waterloo Archives and Special Collections, Waterloo.
121 McGarry, *Ghosts of Futures Past,* 67, 75–6.
122 Merle W. Hersey, *Seventy-Fifth Anniversary of the Lily Dale Assembly, 1879-1954, Lily Dale N.Y.: Condensed History* (Lily Dale, NY: Lily Dale Book Shop, 1954), 7, 51.
123 Seance notes, June 30, 1930, Maines Pincock Family Fonds, GA64, series 4, file 99, University of Waterloo Archives and Special Collections, Waterloo.
124 Ibid.; Huhndorf, *Going Native,* 103.
125 Seance notes, April 6, 1931, Maines Pincock Family Fonds, GA64, series 4, file 102, University of Waterloo Archives and Special Collections, Waterloo.
126 O'Hara Pincock, *Trails of Truth,* 102.
127 Alan Trachtenberg, *Shades of Hiawatha: Staging Indians, Making Americans* (New York: Hill and Wang, 2004), 31, 211.
128 Noel Dyck, "Canadian Anthropology and the Ethnography of 'Indian Administration,'" in *Historicizing Canadian Anthropology,* ed. Julia Harrison and Regna Darnell (Vancouver: UBC Press, 2006), 81.
129 Andrew Nurse, "Marius Barbeau and the Methodology of Salvage Ethnography in Canada, 1911-51," in *Historicizing Canadian Anthropology,* ed. Julia Harrison and Regina Darnell (Vancouver: UBC Press, 2006), 53–55.
130 Trachtenberg, *Shades of Hiawatha,* 31, 211.
131 An Act to Amend the Indian Act, 1914, *Statutes of Canada,* chs. 4–5, 225.
132 George Duncan Campbell Scott, quoted in Celia Haig-Brown, "Resistance and Renewal: First Nations Aboriginal Education in Canada," in *Race and Racialization: Essential Readings,* ed. Tania Das Gupta et al. (Toronto: Canadian Scholars' Press, 2007), 172.
133 Trachtenberg, *Shades of Hiawatha,* xvii-xviii.
134 Philip J. Deloria, *Indians in Unexpected Places* (Lawrence: University Press of Kansas, 2004), 50.
135 Philip J. Deloria, *Playing Indian* (New Haven, CT: Yale University Press, 1998), 120.
136 McComas, *Ghosts I Have Talked With,* 37.
137 "Instructions and Work for a 'Materialization,'" Seance notes, November 3, 1929, Hamilton Family Fonds, MSS14 (A.79-41), box 16, folder 2, University of Manitoba Archives and Special Collections, Winnipeg.
138 Seance notes, October 17, 1929, Maines Pincock Family Fonds, GA64, series 4, file 93, University of Waterloo Library, Winnipeg.
139 Seance notes, June 25, 1932, Maines Pincock Family Fonds, GA64, series 4, file 125, University of Waterloo Library, Waterloo.
140 L. Hamilton, "Steps in the Hamilton Inquiry" (c. 1945), Hamilton Family Fonds, MSS14 (A.79-41), box 15, folder 6, University of Manitoba Archives and Special Collections, Winnipeg.
141 Lillian Hamilton, "Special Sitting, Black Hawk Appears," April 11, 1928, Hamilton Family Fonds, MSS14 (A.79-41), box 15, folder 11, University of Manitoba Archives and Special Collections, Winnipeg.

142 Rosemarie K. Banks, "Staging the 'Native': Making History in American Theatre Culture, 1828–1932," *Theatre Journal* 15, 4 (1993): 463.
143 McGarry, *Ghosts of Futures Past,* 70.
144 S. Elizabeth Bird, "Introduction: Constructing the Indian, 1830s-1990s," in *Dressing in Feathers: The Construction of the Indian in American Popular Culture,* ed. S. Elizabeth Bird (Boulder, CO: Westview, 1996), 2–3.
145 Seance notes, June 4, 1928, Hamilton Family Fonds, MSS14 (A.79-41), box 15, folder 12, University of Manitoba Archives and Special Collections, Winnipeg.
146 Seance notes, September 5, 1928, Hamilton Family Fonds, MSS14 (A.79-41), box 15, folder 12, University of Manitoba Archives and Special Collections, Winnipeg.
147 Seance notes, September 8, 1928, Hamilton Family Fonds, MSS14 (A.79-41), box 15, folder 12, University of Manitoba Archives and Special Collections, Winnipeg.
148 Deloria, *Indians in Unexpected Places,* 49–50.
149 Seance notes, December 23, 1928, Hamilton Family Fonds, MSS14 (A.79-41), box 15, folder 13, University of Manitoba Archives and Special Collections, Winnipeg.
150 "A Costly Mistake," Seance notes, October 20, 1929, Hamilton Family Fonds, MSS14 (A.79-41), box 16, folder 1, University of Manitoba Archives and Special Collections, Winnipeg.
151 "A Light Promised," Seance notes, May 10, 1928, Hamilton Family Fonds, MSS14 (A.79-41), box 15, folder 11, University of Manitoba Archives and Special Collections, Winnipeg.
152 "Black Hawk Foretells Great Work," Seance notes, May 14, 1928, Hamilton Family Fonds, MSS14 (A.79-41), box 15, folder 11, University of Manitoba Archives and Special Collections, Winnipeg.
153 David R. Roediger, *The Wages of Whiteness: Race and the Making of the American Working Class* (1991; reprint, London: Verso, 1996), 21.
154 "Black Hawk Foretells Great Work."
155 Lillian Hamilton, "Vision of the Two Keys," May 20, 1928, Hamilton Family Fonds, MSS14 (A.79-41), box 15, folder 11, University of Manitoba Archives and Special Collections, Winnipeg.

Chapter 3: A Touch of the Uncanny

1 William Crookes, quoted in T.G. Hamilton, "Crookes 'Radiant Matter'" (1929), Hamilton Family Fonds, MSS14 (A.79-41), box 15, folder 10, University of Manitoba Archives and Special Collections, Winnipeg.
2 Marshall McLuhan, "Inside the Five Sense Sensorium," *Canadian Architect* 6 (1961): 49–50.
3 Constance Classen, "McLuhan in the Rainforest: The Sensory Worlds of Oral Cultures," in *Empire of the Senses: The Sensual Culture Reader,* ed. David Howes (Oxford: Berg, 2005), 147–48.
4 Lillian Hamilton, "More Deep Trance Writing," April 3, 1931, Hamilton Family Fonds, MSS14 (A.79-41), box 16, folder 6, University of Manitoba Archives and Special Collections, Winnipeg.
5 Seance notes, January 26, 1930, Maines Pincock Family Fonds, GA64, series 4, file 94, University of Waterloo Library, Waterloo.
6 Seance notes, January 27, 1930, Maines Pincock Family Fonds, GA64, series 4, file 97, University of Waterloo Library, Waterloo.

7 Seance notes, May 11, 1931, Maines Pincock Family Fonds, GA64, series 4, file 109, University of Waterloo Library, Waterloo.
8 Ibid.
9 Ibid.
10 Seance notes, September 25, 1927, Maines Pincock Family Fonds, GA64, series 4, file 79, University of Waterloo Library, Waterloo.
11 J. Malcolm Bird, *Margery the Medium* (Boston: Small, Maynard and Company, 1925), 174; "Apport Received," Seance notes, August 4, 1929, Hamilton Family Fonds, MSS14 (A.79-41), box 15, folder 16, University of Manitoba Archives and Special Collections, Winnipeg.
12 Seance notes, October 7, 1929, Maines Pincock Family Fonds, GA64, series 4, file 93, University of Waterloo Library, Waterloo.
13 Jenny O'Hara Pincock to Minnie Maines, September 6, 1927, Maines Pincock Family Fonds, GA64, series 1, file 2, University of Waterloo Library, Waterloo; Seance notes, July 1, 1928, Maines Pincock Family Fonds, GA64, series 4, file 81, University of Waterloo Library, Waterloo.
14 Seance notes, September 25, 1927, Maines Pincock Family Fonds, GA64, series 4, file 79, University of Waterloo Library, Waterloo.
15 Seance notes, July 1, 1928, Maines Pincock Family Fonds, GA64, series 4, file 81, University of Waterloo Library, Waterloo.
16 Marlene Tromp, *Altered States: Sex, Nation, Drugs, and Self-Transformation in Victorian Spiritualism* (Albany: State University of New York Press, 2006), 11; Jeffrey Sconce, *Haunted Media: Electronic Presence from Telegraphy to Television* (Durham, NC: Duke University Press, 2000), 27.
17 Seance notes, January 27, 1930, Maines Pincock Family Fonds, GA64, series 4, file 97, University of Waterloo Library, Waterloo; T.G. Hamilton, *Intention and Survival: Psychical Research Studies and the Bearing of Intentional Actions by Trance Personalities on the Problem of Human Survival,* ed. J.D. Hamilton (Toronto: Macmillan, 1942), 24.
18 Seance notes, September 9, 1928, Maines Pincock Family Fonds, GA64, series 4, file 86, University of Waterloo Library, Waterloo.
19 Seance notes, January 27, 1930, Maines Pincock Family Fonds, GA64, series 4, file 97, University of Waterloo Library, Waterloo.
20 T.G. Hamilton, "Records II: Miniature Teleplasmic Faces," October 7, 1928, Hamilton Family Fonds MSS14 (A.79-41), box 15, folder 13, University of Manitoba Archives and Special Collections, Winnipeg; "Walter's Preparations Completed," Seance notes, September 17, 1928, Hamilton Family Fonds MSS14 (A.79-41), box 15, folder 12, University of Manitoba Archives and Special Collections, Winnipeg; T.G. Hamilton, "Report of a Sitting Held at 185 Kelvin St., Winnipeg," November 12, 1930, Hamilton Family Fonds, MSS14 (A.79-41), box 16, folder 4, University of Manitoba Archives and Special Collections, Winnipeg; Margaret L. Hamilton, "Third Teleplasmic Structure," Hamilton Family Fonds, MSS14 (A.79-41), box 15, file 12, University of Manitoba Archives and Special Collections, Winnipeg.
21 Bird, *Margery the Medium,* 78, 218–19.
22 "Walter's Preparations Completed."
23 Bird, *Margery the Medium,* 321.
24 Ibid., 292–93.

25 T.G. Hamilton, "The Mediumship of Mary Marshall," Hamilton Family Fonds, MSS14 (A.79-41), box 15, file 12, University of Manitoba Archives and Special Collections, Winnipeg.
26 Leigh Eric Schmidt, *Hearing Things: Religion, Illusion, and the American Enlightenment* (Cambridge, MA: Harvard University Press, 2000), 103–4.
27 Ibid., 109.
28 Anna Louise Fletcher to William Cartheuser, February 15, 1927, Maines Pincock Family Fonds, GA64, series 1, file 2, University of Waterloo Library, Waterloo.
29 Jenny O'Hara Pincock to Minnie Maines, September 9, 1927, Maines Pincock Family Fonds, GA64, series 1, file 2, University of Waterloo Library, Waterloo.
30 Seance notes, January 27, 1930, Maines Pincock Family Fonds, GA64, series 4, file 97, University of Waterloo Library, Waterloo.
31 Henry C. McComas, *Ghosts I Have Talked With* (Baltimore, MD: Williams and Wilkins Company, 1935), 37.
32 Ibid., 51.
33 L. Hamilton, "Talking Machine or Voice Box," July 10, 1929, Hamilton Family Fonds, MSS14 (A.79-41), box 15, folder 16, University of Manitoba Archives and Special Collections, Winnipeg.
34 E.E. Dudley, "Telekinesis of Later 1927," *Proceedings of the American Society for Psychical Research: The Margery Mediumship,* vol. 2 (New York: American Society for Psychical Research, 1933), 670; "Sitting at 10 Lime Street," December 12, 1927, Margery Fonds, box 119, file 8, American Society of Psychical Research Library and Archives, New York; Bird, *Margery the Medium,* 294.
35 B. Wentworth Emmons, "Report of a Sitting with Mrs. L.R.G. Crandon ('Margery')," April 22, 1925, Margery Fonds, box 119, file 2, American Society for Psychical Research Library and Archives, New York.
36 J.B. Rhine, "Sitting at House of Dr. R.G. Crandon, 10 Lime Street," July 1, 1926, Margery Fonds, box 119, file 14, American Society for Psychical Research Library and Archives, New York.
37 David Howes, "Introduction," in *The Empire of the Senses: The Sensual Culture Reader,* ed. David Howes (2005; reprint, Oxford: Berg, 2006), 3.
38 Ibid., 2.
39 Edward Said, *Orientalism* (New York: Vintage Books, 1979), 72.
40 Bird, *Margery the Medium,* 53, 55.
41 Mark A. Richardson, "The Chinese Script and Cross-Correspondences," in *Proceedings of the American Society for Psychical Research: The Margery Mediumship,* vol. 2 (New York: American Society for Psychical Research, 1933), 809-10.
42 Ibid., 807, 810.
43 Ibid., 821.
44 Simon DeBrath, "Chinese Script and Voices," *Psychic Science: Quarterly Transactions of the British College* 8, 3 (1929): 214–16.
45 Seance notes, December 16, 1928, Hamilton Family Fonds, MSS14 (A.79-41), box 15, folder 13, University of Manitoba Archives and Special Collections, Winnipeg.
46 "A Children's Party Held on March 16th, 1929 at 47 Church Street, St. Catharines Ontario," Maines Pincock Family Fonds, GA64, series 4, file 91, University of Waterloo Library, Waterloo.
47 "The Fair Young Man Appears," Seance notes, February 12, 1928, Hamilton Family Fonds, MSS14 (A.79-41), box 15, folder 11, University of Manitoba Archives and Special Collections, Winnipeg.

48 "First Conversation with Walter, First Perfume," Seance notes, March 4, 1928, Hamilton Family Fonds, MSS14 (A.79-41), box 15, file 1, University of Manitoba Archives and Special Collections, Winnipeg.
49 Mark M. Smith, *How Race Is Made: Slavery, Segregation and the Senses* (Chapel Hill: University of North Carolina Press, 2006), 3–4, 9.
50 Lisa Rose Mar, *Brokering Belonging: Chinese in Canada's Exclusion Era, 1885-1995* (Toronto: University of Toronto Press, 2010), 3–4; Erika Lee, *At America's Gates: Chinese Immigration during the Exclusion Era, 1882-1943* (Chapel Hill: University of North Carolina Press, 2007), 2–18; Erika Lee and Judy Yung, *Angel Island: Immigrant Gateway to America* (Oxford: Oxford University Press, 2010), 74–76, 169–72.
51 Marilyn Lake and Henry Reynolds, *Drawing the Global Colour Line: White Men's Countries and the International Challenge of Racial Equality* (Cambridge, UK: Cambridge University Press, 2008), 4–5.
52 Ibid., 6, 313.
53 Homi K. Bhabha, *The Location of Culture* (London: Routledge, 1994), 96.
54 Richard King, *Orientalism and Religion: Postcolonial Theory, India and 'The Mystic East'* (London: Routledge, 1999), 11–12, 33–34.
55 Seance notes, January 6, 1932, Maines Pincock Family Fonds, GA64, series 4, file 119, University of Waterloo Library, Waterloo.
56 Ronald Inden, *Imagining India* (Oxford: Blackwell, 1992), 104, 127.
57 "First Conversation with Walter, First Perfume."
58 E.E. Dudley, "Five Outstanding Records of 1927," in *Proceedings of the American Society for Psychical Research: The Margery Mediumship,* vol. 2 (New York: American Society for Psychical Research, 1933), 617.
59 Mari Yoshihara, *Embracing the East: White Women and American Orientalism* (Oxford: Oxford University Press, 2003), 7–8, 99–100, 125.
60 Ibid., 3–4; Hugh B. Urban, "'India's Darkest Heart': Kali in the Colonial Imagination," in *Encountering Kali: In the Margins, at the Center, in the West,* ed. Rachel Fell McDermott and Jeffrey J. Kripal (Berkeley: University of California Press, 2003), 172–73.
61 Said, *Orientalism,* 44.
62 Inden, *Imagining India,* 127.
63 McComas, *Ghosts I Have Talked With,* 30.
64 "Sitters Have Strong Subjective Sensations," Seance notes, May 23, 1928, Hamilton Family Fonds, MSS14 (A.79-41), box 15, folder 11, University of Manitoba Archives and Special Collections, Winnipeg.
65 E.J. Dingwall, *How to Go to a Medium: A Manual of Instruction* (London: Kegan Paul, Trench, Trubner and Company, 1922), 6.
66 McComas, *Ghosts I Have Talked With,* 30.
67 Seance notes, September 25, 1927, Maines Pincock Family Fonds, GA64, series 4, file 79, University of Waterloo Library, Waterloo; Seance notes, October 17, 1929, Maines Pincock Family Fonds, GA64, series 4, file 93, University of Waterloo Library, Waterloo; Bird, *Margery the Medium,* 178.
68 Constance Classen, David Howe, and Anthony Synnott, *Aroma: The Cultural History of Smell* (London: Routledge, 1994), 84.
69 Ibid., 83.
70 Bird, *Margery the Medium,* 178.
71 Walter Franklin Prince, "Margery Case Notes," Margery Fonds, box 117, file 4, American Society for Psychical Research Library and Archives, New York.

72 J. Malcolm Bird, "A.S.P.R. Seances: Second Series," in *Proceedings of the American Society for Psychical Research: The Margery Mediumship,* vol. 1 (New York: American Society for Psychical Research, 1928), 182; "Sitting of July 17th, 1925, 10 Lime Street, Boston, Mass.," Margery Fonds, box 119, file 5, American Society for Psychical Research Library and Archives, New York.

73 "Sitting at 10 Lime Street," March 6, 1926, Margery Fonds, box 119, file 7, American Society for Psychical Research Library and Archives, New York; J. Malcolm Bird, "Paraffin Gloves under the Co-mediumship of Dr. Hardwicke," in *Proceedings of the American Society for Psychical Research: The Margery Mediumship,* vol. 1 (New York: American Society for Psychical Research, 1928), 454.

74 E.J. Dingwall, "A Report of a Series of Sittings with the Medium Margery, June 1926," in *Proceedings of the Society for Psychical Research,* vol. 36 (Glasgow, UK: Robert MacLehose and Company, 1928), 110, 134, 140.

75 Constance Classen, *The Color of Angels: Cosmology, Gender and the Aesthetic Imagination* (London and New York: Routledge, 1998), 57.

76 Ibid.

77 Hereward Carrington, *The Invisible World* (New York: Beechhurst Press and B. Ackerman, 1946), 75.

78 "Sitting at 10 Lime Street," April 28, 1925, Margery Fonds, box 119, file 5, American Society for Psychical Research Library and Archives, New York; "Sitting at 10 Lime Street," June 11, 1925, Margery Fonds, box 119, file 5, American Society for Psychical Research Library and Archives, New York; J. Malcolm Bird, "The Teleplasmic Hand," in *Proceedings of the American Society for Psychical Research: The Margery Mediumship,* vol. 1 (New York: American Society for Psychical Research, 1928), 38, 46.

79 Constance Classen, *The Deepest Sense: A Cultural History of Touch* (Urbana: University of Illinois Press, 2012), xiv, 75–76; Carolyn Korsmeyer, *Making Sense of Taste: Food and Philosophy* (Ithaca, NY: Cornell University Press, 1999), 21.

80 Seance notes, September 12, 1933, Maines Pincock Family Fonds, GA64, series 4, file 141, University of Waterloo Library, Waterloo.

81 Bird, *Margery the Medium,* 423.

82 Carrington, *Invisible World,* 90.

83 Ibid., 7.

84 Lorraine Daston and Peter Galison, *Objectivity* (New York: Zone Books, 2007), 36.

85 Ibid., 37.

86 T.G. Hamilton to Mr. R. Sproull, April 14, 1931, Hamilton Family Fonds, MSS14 (A.79-41), box 5, folder 1, University of Manitoba Archives and Special Collections, Winnipeg.

87 Seance records, April 28, 1929, Hamilton Family Fonds, MSS14 (A.79-41), box 14, folder 21, University of Manitoba Archives and Special Collections, Winnipeg; Hamilton, *Intention and Survival,* 34.

88 T.G. Hamilton, "Telekinesis" (1927), Hamilton Family Fonds, MSS14 (A.79-41), box 9, folder 13, University of Manitoba Archives and Special Collections, Winnipeg; Hamilton, *Intention and Survival,* 35–36.

89 Bird, *Margery the Medium,* 135.

90 Ibid., 227.

91 Ibid., 179, 225.

92 J. Malcolm Bird, "The Glass Cabinet and the Wired Control," in *Proceedings of the American Society for Psychical Research: The Margery Mediumship,* vol. 1 (New York: American

Society for Psychical Research, 1928), 321; "Sitting at 10 Lime Street, Boston, Mass.," January 30, 1926, Margery Fonds, box 119, file 7, American Society for Psychical Research Library and Archives, New York.

93 Linda Gordon, *Heroes of Their Own Lives: The Politics and History of Family Violence, Boston, 1880-1960* (Urbana: University of Illinois Press, 2002), 250–88.

94 Classen, *Color of Angels,* 6.

95 Bird, *Margery the Medium,* 259.

96 Ibid., 260.

97 Ibid., 425–26, 431.

98 Bird, "Glass Cabinet," 312.

99 Bird, *Margery the Medium,* 475.

100 "Report of Sitting at 10 Lime Street," Seance notes, January 27, 1926, Margery Fonds, box 119, file 7, American Society for Psychical Research Library and Archives, New York; "Sitting at 10 Lime Street," Seance notes, January 28, 1926, Margery Fonds, box 114, file 2, American Society for Psychical Research Library and Archives, New York.

101 J. Malcolm Bird, "The Margery Mediumship: An Experiment in Fraud-Proof Control of a New Type," *Journal of the American Society for Psychical Research* 20, 11 (1926): 676–77.

102 Bird, "Glass Cabinet," 314.

103 William H. Button, "The Margery Mediumship: A Solus Sitting for Thumb Print March 11th, 1931 under Additional Techniques of Control," in *The Margery Mediumship: Records of Supernormal Production of Thumbprints,* ed. R.J. Tillyard (New York: American Society for Psychical Research, 1931), 13.

104 Michel Foucault, *Discipline and Punish: The Birth of the Prison,* trans. Alan Sheridan (New York: Vintage Books, 1977), 135–38.

105 J.B. Rhine and Louisa E. Rhine, "One Evening's Observation on the Margery Mediumship," *Journal of Abnormal and Social Psychology* 21, 4 (1927): 406–7.

106 Bird, *Margery the Medium,* 471–73; "Record of a Séance with Margery and Walter Franklin Prince of October 30th, 1924," Margery Fonds, box 116, file 11, American Society for Psychical Research Library and Archives, New York.

107 "Sitting at 10 Lime Street," January 9, 1925, Margery Fonds, box 119, file 4, American Society for Psychical Research Library and Archives, New York.

108 "Agreement on Procedures of the Society for Psychical Research Commission" (1925), Margery Fonds, box 119, file 4, American Society for Psychical Research Library and Archives, New York.

109 J. Malcolm Bird, "Report to the Research Committee on Sittings with 'Margery' Held from June 10th to October 12th, 1925, with Recommendations for Future Procedures," Margery Fonds, box 119, file 2, American Society for Psychical Research Library and Archives, New York.

110 Seance notes, September 18, 1926, Margery Fonds, box 119, file 7, American Society for Psychical Research Library and Archives, New York.

111 Joseph J. Skirball, quoted in R.J. Tillyard, "Dr. R.J. Tillyard's Notes of His Séance with Margery," in *The Margery Mediumship: Record of Supernormal Production of Thumbprints,* ed. R.J. Tillyard (New York: American Society for Psychical Research, 1931), 11.

112 Evelyn Fox Keller, *Secrets of Life, Secrets of Death: Essays on Language, Gender and Science* (New York: Routledge, 1992), 19. For a similar argument, see Donna Haraway, *Modest_Witness@Second_Millennium.FemaleMan©_Meets_OncoMouse™: Feminism and Technoscience* (New York: Routledge, 1997), 24.

113 Korsmeyer, *Making Sense of Taste,* 3.
114 Classen, *Color of Angels,* 6; Korsmeyer, *Making Sense of Taste,* 31.
115 Classen, *Color of Angels,* 6.
116 Korsmeyer, *Making Sense of Taste,* 21, 25, 31.
117 Bird, *Margery the Medium,* 73, 293–94.
118 Ibid., 294.
119 Ibid., 298–301.
120 Emmons, "Report of a Sitting with Mrs. L.R.G. Crandon."
121 T.G. Hamilton, "Telekinesis: Phenomenal Sounds," January 8, 1922, Hamilton Family Fonds, MSS14 (A.79-41), box 9, folder 13, University of Manitoba Archives and Special Collections, Winnipeg.
122 T.G. Hamilton, "First Whispering Voice Heard," June 2, 1928, Hamilton Family Fonds, MSS14 (A.79-41), box 15, file 12, University of Manitoba Archives and Special Collections, Winnipeg; T.G. Hamilton, "The Voice Improving," June 13, 1928, Hamilton Family Fonds, MSS14 (A.79-41), box 15, file 12, University of Manitoba Archives and Special Collections, Winnipeg.
123 Hamilton, "Telekinesis: Phenomenal Sounds."
124 Bird, *Margery the Medium,* 22.
125 Jenny O'Hara Pincock to "Everybody," October 20, 1927, Maines Pincock Family Fonds, GA64, series 1, file 2, University of Waterloo Library, Waterloo.
126 T.G. Hamilton, "Telekinesis: Teleplasmic Raps Contact and Non-Contact" (1927), Hamilton Family Fonds, MSS14 (A.79-41), box 9, folder 13, University of Manitoba Archives and Special Collections, Winnipeg.
127 Ibid.
128 Ibid.
129 Le Roi Crandon, "Sitting at 10 Lime Street," January 26, 1926, Margery Fonds, box 119, file 7, American Society for Psychical Research Library and Archives, New York.
130 J. Malcolm Bird, "Another Committee," in *Proceedings of the American Society for Psychical Research: The Margery Mediumship,* vol. 1 (New York: American Society for Psychical Research, 1928), 386.
131 Mark M. Smith, *Sensory History* (Oxford: Berg, 2007), 25; Marita Sturken and Lisa Cartwright, *Practices of Looking: An Introduction to Visual Culture* (2001; reprint, Oxford: Oxford University Press, 2003), 299.
132 John Tagg, *The Burden of Representation: Essays on Photographies and Histories* (London: Macmillan Education, 1988), 234.
133 Smith, *Sensory History,* 19–24.
134 Howes, "Introduction," 5.
135 Tromp, *Altered States,* 11–12; Ruth Brandon, *The Spiritualists: The Passion for the Occult in the Nineteenth and Twentieth Centuries* (Buffalo: Prometheus Books, 1984), 102.
136 McComas, *Ghosts I Have Talked With,* 7, 30, 33.
137 William McDougall, "Further Observations on the 'Margery' Case," *Journal of the American Society for Psychical Research* 19, 6 (1925): 298.
138 J. Malcolm Bird, "The Bell-Boxes of Early 1925," in *Proceedings of the American Society for Psychical Research: The Margery Mediumship,* vol. 1 (New York: American Society for Psychical Research, 1928), 115; Dingwall, "Report of a Series," 90.
139 Korsmeyer, *Making Sense of Taste,* 21.
140 Classen, *Color of Angels,* 66; Korsmeyer, *Making Sense of Taste,* 3.
141 Haraway, *Modest_Witness,* 24.

Chapter 4: The Qualities of Quartz

1 Hans Driesch, "Walter Prince as a Scientist," in *Walter Franklin Prince: A Tribute to His Memory,* ed. Boston Society for Psychical Research (Boston: Boston Society for Psychical Research, 1935), 36.

2 John Tresch, *The Romantic Machine: Utopian Science and Technology after Napoleon* (Chicago: University of Chicago Press, 2012), xi.

3 Henry C. McComas, *Ghosts I Have Talked With* (Baltimore, MD: Williams and Wilkins Company, 1935), 3.

4 Peter J. Bowler, *Science for All: The Popularization of Science in Early Twentieth-Century Britain* (Chicago: University of Chicago Press, 2009), 21–25; Robert L. O'Connell, *Of Arms and Men: A History of War, Weapons and Aggression* (Oxford: Oxford University Press, 1989), 242, 269.

5 Glen Scott Allen, *Master Mechanics and Wicked Wizards: Images of the American Scientist as Hero and Villain from Colonial Times to the Present* (Amherst: University of Massachusetts Press, 2009), 68.

6 Miles Orvell, *The Real Thing: Imitation and Authenticity in American Culture, 1880-1940* (Chapel Hill: University of North Carolina Press, 1989), 157; Stanley J. Reiser, "Technology and the Use of the Senses in Twentieth-Century Medicine," in *Medicine and the Five Senses,* ed. W.F. Bynum and Roy Porter (Cambridge, UK: Cambridge University Press, 1993), 264; Daniel Patrick Thurs, *Science Talk: Changing Notions of Science in American Popular Culture* (New Brunswick, NJ: Rutgers University Press, 2007), 96, 119; Juliann Sivulka, *Soap, Sex and Cigarettes: A Cultural History of American Advertising,* 2nd ed. (Boston: Wadworth, 2012), 125.

7 E.J. Dingwall, *How to Go to a Medium: A Manual of Instruction* (London: Kegan Paul, Trench, Trubner and Company, 1922), 20.

8 McComas, *Ghosts I Have Talked With,* 3.

9 Lorraine Daston and Peter Galison, *Objectivity* (New York: Zone Books, 2007), 311.

10 On the historical emergence of the stethoscope, see Stanley J. Reiser, *Medicine and the Reign of Technology* (Cambridge, UK: Cambridge University Press, 1978), 23–29; Malcolm Nicolson, "The Introduction of Percussion and Stethoscopy to Early Nineteenth-Century Edinburgh," in *Medicine and the Five Senses,* ed. W.F. Bynum and Roy Porter, 134–53 (Cambridge, UK: Cambridge University Press, 1993); and Leigh Eric Schmidt, *Hearing Things: Religion, Illusion and the American Enlightenment* (Cambridge, MA: Harvard University Press, 2000), 1–3.

11 Reiser, *Medicine and the Reign,* 38.

12 Hereward Carrington, *The Invisible World* (New York: Beechhurst Press and B. Ackerman, 1946), 81.

13 McComas, *Ghosts I Have Talked With,* 37.

14 Ibid., 50.

15 Ibid., 51. Hereward Carrington tells of how he attempted to try this test on Cartheuser. The medium, however, refused to permit Carrington to use the stethoscope and thus to repeat McComas's test. Carrington, *Invisible World,* 81.

16 J. Malcolm Bird, *Margery the Medium* (Boston: Small, Maynard and Company, 1925), 272.

17 Weighing experiments of this nature were reported in Bird, *Margery the Medium,* 272–85; J. Malcolm Bird, "The Rehabilitation of the Scales," in *Proceedings of the American Society for Psychical Research: The Margery Mediumship,* vol. 1 (New York: American Society for

Psychical Research, 1928), 417–27; E.E. Dudley, "Telekinesis in 1928 and 1929," in *Proceedings of the American Society for Psychical Research: The Margery Mediumship,* vol. 2, 672–98 (New York: American Society for Psychical Research, 1933); E.E. Dudley, "Photographic Experiments in 1927-8," in *Proceedings of the American Society for Psychical Research: The Margery Mediumship,* vol. 2, 699–714 (New York: American Society for Psychical Research, 1933).

18 T.G. Hamilton, "Telekinesis: Alteration of Table Weight," Hamilton Family Fonds, MSS14 (A.79-41), box 9, folder 13, University of Manitoba Archives and Special Collections, Winnipeg.

19 Ibid.

20 Bird, *Margery the Medium,* 78–80; Seance notes, April 3, 1925, Margery Fonds, box 119, file 6, American Society for Psychical Research Library and Archives, New York.

21 Bird, *Margery the Medium,* 301–2; J. Malcolm Bird, "Informal Tests of the Walter Voice," in *Proceedings of the American Society for Psychical Research: The Margery Mediumship,* vol. 1 (New York: American Society for Psychical Research, 1928), 25–30; Seance notes, March 31, 1925, Margery Fonds, box 119, file 6, American Society for Psychical Research Library and Archives, New York.

22 This contraption is described in Seance notes, May 21, 1925, Margery Fonds, box 119, file 6, American Society for Psychical Research Library and Archives, New York; and J. Malcolm Bird, "Dr. Richardson's Voice-Control Machine," in *Proceedings of the American Society for Psychical Research: The Margery Mediumship,* vol. 1 (New York: American Society for Psychical Research, 1928), 258–59.

23 Bird, "Dr. Richardson's Voice-Control Machine," 262–64.

24 Ibid., 268.

25 Ibid., 275; Lillian Hamilton, "Teleplasmic Researchers Carried Out by T. Glen Hamilton, Records" (1928), Hamilton Family Fonds, MSS14 (A.79-41), box 15, folder 11, University of Manitoba Archives and Special Collection, Winnipeg.

26 Bird, *Margery the Medium,* 153–54, commends Comstock for the use of his equipment.

27 Ibid., 150, 260.

28 Ibid., 287-88.

29 Ibid., 392–97.

30 T.G. Hamilton, "Teleplasmic Phenomenon in Winnipeg," *Psychic Science: Quarterly Transactions of the British College* 8, 3 (1929): 187.

31 T.G. Hamilton, "Margery in Winnipeg: Three Séances of December 1926," in *Proceedings of the American Society for Psychical Research: The Margery Mediumship,* vol. 2 (New York: American Society of Psychical Research, 1933), 558.

32 T.G. Hamilton, "Notes on the Walter Control," Hamilton Family Fonds, MSS14 (A.79-41), box 14, folder 21, University of Manitoba Archives and Special Collections, Winnipeg.

33 Hamilton, "Teleplasmic Phenomenon in Winnipeg," 185.

34 Bird, *Margery the Medium,* 426.

35 John Tagg, *The Burden of Representation: Essays on Photographies and Histories* (London: Macmillan Education, 1988), 3–5; Suren Lalvani, *Photography, Vision and the Production of Modern Bodies* (New York: State University of New York Press, 1996), 2–3; Jonathan Crary, *Techniques of the Observer: On Vision and Modernity in the Nineteenth Century* (1990; reprint, Cambridge, MA: MIT Press, 1995), 8–9.

36 Reiser, "Technology and the Use," 262.

37 Orvell, *Real Thing,* xviii; Stephen Kern, *The Culture of Time and Space, 1880-1918* (Cambridge, MA: Harvard University Press, 1983), 1–2, 7; Reiser, "Technology and the Use," 262–66.
38 T.G. Hamilton to Harvey Agnew, March 1928, Hamilton Family Fonds, MSS14 (A.79-41), box 4, file 3, University of Manitoba Archives and Special Collections, Winnipeg.
39 Irving R. Gaertner, "The Acting Power of Light as a Disturbing Element in Physical Mediumship," *Psychic Research* 25, 1 (1931): 3-4.
40 Jenny O'Hara Pincock, *Trails of Truth* (Los Angeles, CA: Austin, 1930), 21.
41 Carrington, *Invisible World,* 30.
42 O'Hara Pincock, *Trails of Truth,* 21.
43 "Statement of Daniel F. Comstock," Margery Fonds, box 116, file 11, American Society for Psychical Research Library and Archives, New York.
44 Ibid; E.J. Dingwall, "A Report of a Series of Sittings with the Medium Margery, June 1926," in *Proceedings of the Society for Psychical Research,* vol. 36 (Glasgow, UK: Robert MacLehose and Company, 1928), 92.
45 Dingwall, *How to Go to a Medium,* 14.
46 Carrington, *Invisible World,* 75–76.
47 Dingwall, "Report of a Series," 95.
48 McComas, *Ghosts I Have Talked With,* 18; T.G. Hamilton, *Intention and Survival: Psychical Research Studies and the Bearing of Intentional Actions by Trance Personalities on the Problem of Human Survival,* ed. J.D. Hamilton (Toronto: Macmillan, 1942), 9–10.
49 For examples in which ultraviolet light via a mercury lamp was employed in the seance, see McComas, *Ghosts I Have Talked With,* 18–23; and Bird, *Margery the Medium,* 262, 358–60.
50 Carrington, *Invisible World,* 87.
51 Ibid., 86-87.
52 T.G. Hamilton, "Photography of Teleplasm," Hamilton Family Fonds, MSS14 (A.79-41), box 16, folder 1, University of Manitoba Archives and Special Collections, Winnipeg.
53 Bird, *Margery the Medium,* 397.
54 Ibid., 394.
55 Seance notes, August 17, 1924, Margery Fonds, box 119, file 3, American Society for Psychical Research Library and Archives, New York; Bird, *Margery the* Medium, 478–79.
56 Seance notes, November 9, 1924, quoted in Bird, *Margery the Medium,* 478.
57 Dingwall, "Report of a Series," 92.
58 J.B. Rhine and Louisa E. Rhine, "One Evening's Observation on the Margery Mediumship," *Journal of Abnormal and Social Psychology* 21, 4 (1927): 408–9.
59 This was the general form taken in the seances studied here, yet the precise construction of a given seance cabinet could vary considerably. The point was that the medium needed to be entirely enclosed, away from the sitters. Ruth Brandon, *The Spiritualists: The Passion for the Occult in the Nineteenth and Twentieth Centuries* (Buffalo: Prometheus Books, 1984), 102.
60 T.G. Hamilton, "Step of Trance: RLS Vision" (1927), Hamilton Family Fonds, PC12 (79–41), box 9, folder 2, University of Manitoba Archives and Special Collections, Winnipeg.
61 J. Malcolm Bird, "The Glass Cabinet and the Wired Control," in *Proceedings of the American Society for Psychical Research: The Margery Mediumship,* vol. 1 (New York: American Society for Psychical Research, 1928), 314, 319.

62 Ibid., 315.

63 Ibid.

64 Ibid., 316–17.

65 "Sittings of the Commission of Inquiry from the A.S.P.R. (January 26-30th, 1926)," Margery Fonds, box 119, file 7, American Society for Psychical Research Library and Archives, New York; J. Malcolm Bird, "The Bell-Boxes of Early 1925," in *Proceedings of the American Society for Psychical Research: The Margery Mediumship,* vol. 1 (New York: American Society for Psychical Research, 1928), 117; J. Malcolm Bird, "The Curious History of the 'Hole Bell-Box,'" in *Proceedings of the American Society for Psychical Research: The Margery Mediumship,* vol. 1 (New York: American Society for Psychical Research, 1928), 123–24.

66 E.E. Dudley, "The Butler Apparatus," in *Proceedings of the American Society for Psychical Research: The Margery Mediumship,* vol. 2 (New York: American Society for Psychical Research, 1933), 588–93.

67 E.E. Dudley, "Telekinesis of Early 1927," in *Proceedings of the American Society for Psychical Research: Margery Mediumship,* vol. 2 (New York: American Society for Psychical Research, 1933), 578; Dudley, "Butler Apparatus," 588.

68 Jeffrey Sconce, *Haunted Media: Electronic Presence from Telegraphy to Television* (Durham, NC: Duke University Press, 2000), 44; Jill Galvan, *The Sympathetic Medium: Feminine Channelling, the Occult and Communication Technologies, 1859-1919* (Ithaca, NY: Cornell University Press, 2010), 11.

69 Dudley, "Butler Apparatus," 589–90.

70 Ibid., 598–601.

71 Hamilton, *Intention and Survival,* 10.

72 Lalvani, *Photography, Vision,* 169.

73 John Tagg, *The Disciplinary Frame: Photographic Truths and the Capture of Meaning* (Minneapolis: University of Minnesota Press, 2009), xvi, 5–6, 225–26.

74 Tagg, *Burden of Representation,* 5.

75 Jennifer Tucker, *Nature Exposed: Photography as Eye Witness in Victorian Science* (Baltimore, MD: John Hopkins University Press, 2005), 4, 159–93.

76 Lucy Traverse, "L'Âme Hu(main)e: Digital Effluvia, Vital Energies, and the Onanistic Occult," *Nineteenth-Century Contexts* 36, 5 (2014): 535–36; John Harvey, *Photography and Spirit* (London: Reaktion Books, 2007), 27.

77 Louis Kaplan, *The Strange Case of William Mumler, Spirit Photographer* (Minneapolis: University of Minnesota Press, 2008), 30–34; Molly McGarry, *Ghosts of Futures Past: Spiritualism and the Cultural Politics of Nineteenth-Century America* (Berkeley: University of California Press, 2008), 110.

78 Orvell, *Real Thing,* 77.

79 McGarry, *Ghosts of Futures Past,* 110.

80 Orvell, *Real Thing,* 198–99.

81 Edward Weston, quoted in ibid., 220.

82 Marina Warner, *Phantasmagoria: Spirit Visions, Metaphors, and Media into the Twenty-First Century* (Oxford: Oxford University Press, 2006), 222–23, 242, 253.

83 A famous example of one fraudulent twentieth-century spirit photographer was William Hope. Although Oliver Lodge suspected him from the start of their acquaintance, Hope was able to fool William Crookes in 1916, who was convinced Hope had captured an image of his dead wife. Janet Oppenheim, *The Other World: Spiritualism and Psychical Research in England, 1850-1914* (Cambridge, UK: Cambridge University Press, 1985), 71, 351–52.

84 Bird, *Margery the Medium,* 359.
85 Ibid., 361–63.
86 An example is Dingwall, "Report of a Series," 122–26.
87 T.G. Hamilton, "Some New Facts Regarding Teleplasms (an Address Delivered by Hamilton to Members and Friends of the British Medical Association on August 27th, 1930)," *Psychic Science: Quarterly Transactions of the British College* 9, 4 (1931): 263.
88 Ibid., 266.
89 T.G. Hamilton, "Teleplasmic Phenomena in Winnipeg, Article 3: Mary M. Teleplasms, Including Another Miniature Face," *Psychic Science: Quarterly Transactions of the British College* 9, 2 (1930): 89.
90 Hamilton, "Photography of Teleplasm."
91 Ibid.
92 Ibid.
93 Ibid.
94 Hamilton, "Teleplasmic Phenomena in Winnipeg, Article 3," 89.
95 Hamilton, "Photography of Teleplasm."
96 Hamilton, "Teleplasmic Phenomena in Winnipeg, Article 3," 89.
97 T.G. Hamilton to Francis A. Hilton, March 15, 1934. Hamilton Family Fonds, MSS14 (A.79-41), box 5, folder 1, University of Manitoba Archives and Special Collections, Winnipeg.
98 Hamilton, "Photography of Teleplasm."
99 T.G. Hamilton, "Directory of Predictions," November 29, 1931, Hamilton Family Fonds, PC12 (A.79-41), box 8, folder 3, University of Manitoba Archives and Special Collections, Winnipeg.
100 Bruno Latour, *Science in Action: How to Follow Scientists and Engineers through Society* (Milton Keynes, UK: Open University Press, 1987), 68.
101 Ibid., 65, 68–69.
102 Ibid.
103 Bird, *Margery the Medium,* 157–60.
104 T.G. Hamilton, "Analytical Study of E.M. Trances" (1927), Hamilton Family Fonds, MSS14 (A.79-41), box 9, folder 9, University of Manitoba, Archives and Special Collections, Winnipeg.
105 T.G. Hamilton, "Growth of the Poole Phenomena from Simple to More Complex, 1923-1927," Hamilton Family Fonds, MSS14 (A.79-41), box 9, folder 9, University of Manitoba Archives and Special Collection, Winnipeg; T.G. Hamilton, "Notes on the Trance Condition," Hamilton Family Fonds, MSS14 (A.79-41), box 9, folder 11, University of Manitoba Archives and Special Collections, Winnipeg.
106 Mary Poovey, *A History of the Modern Fact: Problems of Knowledge in the Sciences of Wealth and Society* (Chicago: University of Chicago Press, 1998), 4.
107 Ibid., 61.
108 T.G. Hamilton, "Telekinesis" (1928), Hamilton Family Fonds, MSS14 (A.79-41), box 9, folder 13, University of Manitoba Archives and Special Collections, Winnipeg.
109 T.G. Hamilton, "Levitations by Psychic Force (Telekinesis)," Hamilton Family Fonds, PC12 (A.79-41), box 8, folder 1, University of Manitoba Archives and Special Collections, Winnipeg.
110 Clive Scott, *The Spoken Image: Photography and Language* (London: Reaktion Books, 1999), 49.

111 A similar argument is made by James Opp in his analysis of Yousuf Karsh's Cold War photography. James Opp, "Picturing Communism: Yousuf Karsh, Canadair, and Cold War Advertising," in *The Cultural Work of Photography in Canada,* ed. Andrea Kunard and Carol Payne, 120–35 (Montreal and Kingston: McGill-Queen's University Press, 2011).

112 Tucker, *Nature Exposed,* 3–10.

113 Scott, *Spoken Image,* 47.

114 Bruno Latour, *Pandora's Hope: Essays in the Reality of Science Studies* (Cambridge, MA: Harvard University Press, 1999), 56.

115 Ibid., 100–6.

Chapter 5: Fragments of a Spectral Self

1 Simon DeBrath, "Editorial Notes," *Psychic Science: Quarterly Transactions of the British College* 9, 4 (1931): 238.

2 Henry Clay McComas, *Ghosts I Have Talked With* (Baltimore, MD: Williams and Wilkins Company, 1935), 30.

3 T.G. Hamilton, "E.M.'s Attitude toward Her Trance Personalities," Hamilton Family Fonds, MSS14 (A.79-41), box 9, folder 9, University of Manitoba Archives and Special Collections, Winnipeg.

4 Ann Taves, *Fits, Trances and Visions: Experiencing Religion and Explaining Experience from Wesley to James* (Princeton, NJ: Princeton University Press, 1999), 207, 248; John Warne Monroe, *Laboratories of Faith: Mesmerism, Spiritism and Occultism in Modern France* (Ithaca, NY: Cornell University Press, 2008), 11; Sofie Lachapelle, *Investigating the Supernatural: From Spiritism and Occultism to Psychical Research and Metaphysics in France, 1853-1931* (Baltimore, MD: John Hopkins University Press, 2011), 5; Deborah Blum, *Ghost Hunters: William James and the Search for Scientific Proof of Life after Death* (New York: Penguin, 2006), 86.

5 Eugene Taylor, *William James on Consciousness beyond the Margin* (Princeton, NJ: Princeton University Press, 1996), 5–6; F.X. Charet, *Spiritualism and the Foundations of C.G. Jung's Psychology* (New York: State University of New York Press, 1993), 149–63.

6 Pamela Klassen, *Spirits of Protestantism: Medicine, Healing and Liberal Christianity* (Berkeley: University of California Press, 2011), 45–46, 78–82.

7 Elwood Worcester, "Dr. Prince as a Psychic Researcher and as a Psychiatrist," in *Walter Franklin Prince: A Tribute to His Memory,* ed. Boston Society for Psychical Research (Boston: Boston Society for Psychical Research, 1935), 10.

8 Graham Richards, "Britain on the Couch: The Popularization of Psychoanalysis in Britain, 1918-40," *Science in Context* 13, 2 (2000): 189; David Lomas, *The Haunted Self: Surrealism, Psychoanalysis and Subjectivity* (New Haven, CT: Yale University Press, 2000), 1–2; Steven C. Ward, *Modernizing the Mind: Psychological Knowledge and the Remaking of Society* (Westport, CT: Praeger, 2002), 148; Nikolas Rose, *Governing the Soul: The Shaping of the Private Self,* 2nd ed. (London: Free Association Books, 1999), viii.

9 Ian A.M. Nicholson, "Gordon Allport, Character and the 'Culture of Personality,' 1897–1937," in *Evolving Perspectives on the History of Psychology,* ed. Wade E. Pickren and Donald A. Dewsbury (Washington, DC: American Psychological Association, 2002), 325.

10 Working under the tutelage of Paul Eugen Bleuler, Jung became allied with word association tests due to his employment of them in 1901 at the Burghölzli Hospital in Zurich. Lisa Appignanesi, *Mad, Bad and Sad: A History of Women and the Mind Doctors* (2007; reprint, New York: W.W. Norton and Company, 2008), 207–8.

11 Ibid.
12 One such contemporary was psychoanalyst and physician M.D. Eder, who translated and published a collection of these word association tests. Carl Jung, ed., *Studies in Word-Association: Experiments in the Diagnosis of Psychopathological Conditions Carried Out at the Psychiatric Clinic of the University of Zurich under the Direction of C.G. Jung, M.D., L.L.D.*, trans. M.D. Eder (New York: Moffat, Yard and Company, 1919), v-vi. These tests continued to influence developing theories of personality throughout the first half of the twentieth century and became a popular method of measuring individual traits. Ian A.M. Nicholson, *Inventing Personality: Gordon Allport and the Science of Selfhood* (Washington, DC: American Psychological Association, 2003), 91.
13 Simon DeBrath, "Editorial Notes," *Psychic Science: Quarterly Transactions of the British College* 12, 4 (1934): 238–39.
14 Hereward Carrington, quoted in ibid., 239–41.
15 Ibid.
16 Hereward Carrington, *The Invisible World* (New York: Beechhurst Press and B. Ackerman, 1946), 61–62.
17 Ibid., 62.
18 Hereward Carrington, quoted in DeBrath, "Editorial Notes" (1934), 240.
19 Carrington, *Invisible World,* 107.
20 Nikolas Rose, *Inventing Our Selves: Psychology, Power and Personhood* (Cambridge: Cambridge University Press, 1998), 74.
21 Carrington, *Invisible World,* 147.
22 Pamela Thurschwell, *Literature, Technology and Magical Thinking, 1880-1920* (Cambridge, UK: Cambridge University Press, 2001), 18; Ian Hacking, *Rewriting the Soul: Multiple Personality and the Sciences of Memory* (Princeton, NJ: Princeton University Press, 1995), 136.
23 Blum, *Ghost Hunters,* 211.
24 Taves, *Fits, Trances and Visions,* 258.
25 T.G. Hamilton, "General Summary of Poole's Development" (1922–27), Hamilton Family Fonds, MSS14 (A.79-41), box 9, folder 10, University of Manitoba Archives and Special Collections, Winnipeg; L. Hamilton, "Steps in the Hamilton Inquiry" (c. 1945), Hamilton Family Fonds, MSS14 (A.79-41), box 15, folder 6, University of Manitoba Archives and Special Collections, Winnipeg.
26 T.G. Hamilton, "The Mysteries of Teleplasm," Hamilton Family Fonds, MSS14 (A.79-41), box 15, folder 6, University of Manitoba Archives and Special Collections, Winnipeg.
27 T.G. Hamilton, "Robert Louis Stevenson and Elizabeth Poole (E.M.)," Hamilton Family Fonds, MSS14 (A.79-41), box 9, folder 8, University of Manitoba Archives and Special Collections, Winnipeg.
28 T.G. Hamilton, "Notes on the Trance Condition," Hamilton Family Fonds, MSS14 (A.79-41), box 9, folder 11, University of Manitoba Archives and Special Collections, Winnipeg.
29 Hamilton, "E.M.'s Attitude."
30 Rose, *Inventing Our Selves,* 80.
31 Worcester, "Dr. Prince," 10.
32 J. Malcolm Bird, "The Code-Hoagland Episode," in *Proceedings of the American Society for Psychical Research: The Margery Mediumship,* vol. 1 (New York: American Society for Psychical Research, 1928), 91.

33 Jeffrey J. Kripal, *Authors of the Impossible: The Paranormal and the Sacred* (Chicago: University of Chicago Press, 2010), 63.

34 Gardner Murphy, "Dr. Prince and the Dora Case," in *Walter Franklin Prince: A Tribute to His Memory,* ed. Boston Society for Psychical Research (Boston: Boston Society for Psychical Research, 1935), 76.

35 André LeBlanc, "The Origins of the Concept of Dissociation: Paul Janet, His Nephew Pierre, and the Problem of Post-hypnotic Suggestion," *History of Science* 39, 1 (2001): 60–63; Robert W. Rieber, *The Bifurcation of the Self: The History and Theory of Dissociation and Its Disorders* (New York: Springer, 2006), 21–22.

36 LeBlanc, "Origins of the Concept," 62–63; Lachapelle, *Investigating the Supernatural,* 71–72.

37 Hamilton, "E.M.'s Attitude."

38 J. Malcolm Bird, *Margery the Medium* (Boston: Small, Maynard and Company, 1925), 26.

39 Ibid., 49.

40 Carrington, *Invisible World,* 69.

41 Carrington's assessment paralleled Henry Clay McComas's estimation of Cartheuser's mediumship in 1935. McComas, *Ghosts I Have Talked With,* 31-53.

42 Ibid., 71.

43 Ibid., 72, emphasis in original.

44 Seance notes, December 5, 1933, Maines Pincock Family Fonds, GA64, series 4, file 143, University of Waterloo Archives and Special Collections, Waterloo.

45 T.G. Hamilton, "A Study of the Winnipeg Group Mediumship and Its Relation to the Dawn Teleplasms," *Psychic Research* 29, 2 (1934): 117.

46 Hamilton, "Notes on the Trance Condition." This manner of representing disease is described in Sander Gilman, *Difference and Pathology: Stereotypes of Sexuality, Race and Madness* (Ithaca, NY: Cornell University Press, 1985), 24.

47 Ibid.

48 "Hamilton, "Notes on the Trance Condition."

49 Ibid.; Hamilton, "E.M.'s Attitude."

50 L. Hamilton, "Steps in the Hamilton Inquiry."

51 Kripal, *Authors of the Impossible,* 66–67, emphasis in original.

52 Wendy Mitchinson, *The Nature of Their Bodies: Women and Their Doctors in Victorian Canada* (Toronto: University of Toronto Press, 1991), 77, 312–14; Wendy Mitchinson, *Giving Birth in Canada, 1900-1950* (Toronto: University of Toronto Press, 2002), 106; Helen King, "Once upon a Text: Hysteria from Hippocrates," in *Hysteria beyond Freud,* ed. Sander Gilman, Helen King, Roy Porter, and Elaine Showalter (Berkeley: University of California Press, 1993), 13; Elaine Showalter, "Hysteria, Feminism and Gender," in *Hysteria beyond Freud,* ed. Sander Gilman, Helen King, Roy Porter, and Elaine Showalter (Berkeley: University of California Press, 1993), 286, 290.

53 Elaine Showalter, *The Female Malady: Women, Madness and English Culture, 1830-1980* (New York: Pantheon Books, 1985), 3–4.

54 E.J. Dingwall, *How to Go to a Medium: A Manual of Instruction* (London: Kegan Paul, Trench, Trubner and Company, 1922), 5–6.

55 E.J. Dingwall, "A Report of a Series of Sittings with the Medium Margery, June 1926," in *Proceedings of the Society for Psychical Research,* vol. 36 (Glasgow, UK: Robert MacLehose and Company, 1928), 80-84, 93, 141–43.

56 Bird, "Code-Hoagland Episode," 91.

57 T.G. Hamilton, *Intention and Survival: Psychical Research Studies and the Bearing of Intentional Actions by Trance Personalities on the Problem of Human Survival,* ed. J.D. Hamilton (Toronto: Macmillan, 1942), 9.
58 Waldemar Kaempffert, "The Scientist," *Journal of the American Society for Psychical Research* 14, 10 (1920): 486.
59 Showalter, *Female Malady,* 3, 204.
60 Michel Foucault, *The Birth of the Clinic: An Archaeology of Medical Perception,* trans. A.M. Sheridan (1963; reprint, London: Routledge, 2008), 1; Rose, *Governing the Soul,* 138.
61 James Opp, *The Lord for the Body: Religion, Medicine and Protestant Faith Healing in Canada, 1880-1930* (Montreal and Kingston: McGill-Queen's University Press, 2005), 187, 202.
62 Hamilton, *Intention and Survival,* 9.
63 Jenny O'Hara Pincock, *Trails of Truth* (Los Angeles, CA: Austin, 1930), 20.
64 Hamilton, *Intention and Survival,* 9.
65 T.G. Hamilton, "Some New Facts Regarding Teleplasms (an Address Delivered by Hamilton to Members and Friends of the British Medical Association on August 27th, 1930)," *Psychic Science: Quarterly Transactions of the British College* 9, 4 (1931): 269.
66 T.G. Hamilton, "Analytical Study of E.M. Trance," Hamilton Family Fonds, MSS14 (A.79-41), box 9, folder 9, University of Manitoba Archives and Special Collections, Winnipeg.
67 Hamilton, "Study of the Winnipeg Group," 117, 119.
68 Hamilton, "E.M.'s Attitude."
69 Ibid.
70 Jacalyn Duffin, *History of Medicine: A Scandalously Short Introduction,* 2nd ed. (Toronto: University of Toronto Press, 2010), 141.
71 Hamilton, "E.M.'s Attitude."
72 T.G. Hamilton, "Sitting of February 12th, 1931 (#218)," Hamilton Family Fonds, MSS14 (A.79-41), box 16, folder 5, University of Manitoba Archives and Special Collections, Winnipeg; Hamilton, *Intention and Survival,* 30.
73 Hamilton, "Study of the Winnipeg Group," 118–19.
74 His medical examination on this night is recorded in T.G. Hamilton, "Katie King Manifestations in the Mary M. Experiments," *Psychic Science: Quarterly Transactions of the British College* 11, 4 (1933): 267–68.
75 Ibid., 268.
76 T.G. Hamilton, "Division VII: A Mass Appears apart from MM's Body - The Great Lucy Materialization Makes Its Appearance," Hamilton Family Fonds, MSS14 (A.79-41), box 15, folder 10, University of Manitoba Archives and Special Collections, Winnipeg.
77 Ibid.; T.G. Hamilton, "Report of a Sitting Held at 185 Kelvin St., Winnipeg," March 10, 1931, Hamilton Family Fonds, MSS14 (A.79-41), box 16, folder 2, University of Manitoba Archives and Special Collections, Winnipeg.
78 Hamilton, "Division VII."
79 Alex Owen, *The Darkened Room: Women, Power, and Spiritualism in Late Nineteenth Century England* (Chicago: University of Chicago Press, 1989), 45.
80 Ibid., 47–48.
81 Molly McGarry, *Ghosts of Futures Past: Spiritualism and the Cultural Politics of Nineteenth-Century America* (Berkeley: University of California Press, 2008), 106; Diana Basham, *The Trial of Woman: Feminism and the Occult Sciences in Victorian Literature* (Basingstoke, UK: Macmillan, 1992), 182.

82 T.G. Hamilton, "Division 13: The Final Katie Phenomena Appear - The Third Katie Face and Veil, Jan. 3, 1933, and the Katie 'Shell,'" April 23, 1933, Hamilton Family Fonds, MSS14 (A.79-41), box 15, folder 10, University of Manitoba Archives and Special Collections, Winnipeg.
83 Seance notes, February 1, 1933, Hamilton Family Fonds, MSS14 (A.79-41), box 16, folder 11, University of Manitoba Archives and Special Collections, Winnipeg.
84 Hamilton, "Division 13."
85 T.G. Hamilton, "The Shell of Katie Appears, Sitting 329," April 23, 1933, Hamilton Family Fonds, MSS14 (A.79-41), box 16, folder 11, University of Manitoba Archives and Special Collections, Winnipeg.
86 Ibid.
87 Foucault, *Birth of the Clinic,* 161.
88 Ibid., 174–77.
89 Hamilton, "E.M.'s Attitude."
90 DeBrath, "Editorial Notes" (1934), 236.

Chapter 6: Teleplasmic Mechanics

1 Carl Jung, *Modern Man in Search of a Soul,* trans. W.S. Dell and Cary F. Baynes (New York: Harcourt, Brace and Company, 1933), 147–48.
2 Bruno Latour, *Reassembling the Social: An Introduction to Actor-Network Theory* (2005; reprint, Oxford: Oxford University Press, 2007), 72.
3 T.G. Hamilton, "General Summary of Poole's Development" (1934–5), Hamilton Family Fonds, MSS14 (A.79-41), box 9, folder 13, University of Manitoba Archives and Special Collections, Winnipeg.
4 E.E. Dudley, "Teleplasmic Structures and Telekinesis," in *Proceedings of the American Society for Psychical Research: The Margery Mediumship,* vol. 2 (New York: American Society for Psychical Research, 1933), 626; Jenny O'Hara Pincock, *Trails of Truth* (Los Angeles, CA: Austin, 1930), 20.
5 Robert Michael Brain, "Materializing the Medium: Ectoplasm and the Quest for Supranormal Biology in Fin-de-Siècle Science and Art," in *Vibratory Modernism,* ed. Anthony Enns and Shelley Trower (Houndsmills, UK: Palgrave Macmillan, 2013), 115–16.
6 T.G. Hamilton to Harvey Agnew, March 1931, Hamilton Family Fonds, MSS14 (A.79-41), box 4, folder 3, University of Manitoba Archives and Special Collections, Winnipeg.
7 Harvey Agnew to T.G. Hamilton, April 7, 1931, Hamilton Family Fonds, MSS14 (A.79-41), box 4, folder 3, University of Manitoba Archives and Special Collections, Winnipeg.
8 T.G. Hamilton, "Article 3: Teleplasm Found in Psychic Study" (c. 1930), Hamilton Family Fonds, MSS14 (A.79-41), box 15, folder 6, University of Manitoba Archives and Special Collections, Winnipeg.
9 "Sitting at 10 Lime Street," February 24, 1925, Margery Fonds, box 119, file 5, American Society for Psychical Research Library and Archives, New York.
10 Agnew to Hamilton, April 7, 1931.
11 Hamilton, "Article 3."
12 T.G. Hamilton, "Telekinesis: Force Intelligently Applied in a Third Way," Hamilton Family Fonds, MSS14 (A.79-41), box 9, folder 13, University of Manitoba Archives and Special Collections, Winnipeg.
13 Hamilton, "General Summary of Poole's Development."

14 O'Hara Pincock, *Trails of Truth,* 16.
15 Dudley, "Teleplasmic Structures and Telekinesis," 627.
16 O'Hara Pincock, *Trails of Truth,* 20.
17 Patricia Spyer, "Introduction," in *Border Fetishisms: Material Objects in Unstable Places,* ed. Patricia Spyer (New York: Routledge, 1998), 5.
18 Evelyn Fox Keller, *Reflections on Gender and Science* (New Haven, CT: Yale University Press, 1985), 70.
19 J. Malcolm Bird, *Margery the Medium* (Boston: Small, Maynard and Company, 1926), 195–98.
20 These initial experiments are recounted in ibid., 335–43.
21 E.J. Dingwall, "A Report of a Series of Sittings with the Medium Margery, June 1926," in *Proceedings of the Society for Psychical Research,* vol. 36 (Glasgow, UK: Robert MacLehose and Company, 1928), 98.
22 Ibid., 80, 86-98.
23 "Sitting at 10 Lime Street, Fifth Séance," January 6, 1925, Margery Fonds, box 119, file 4, American Society for Psychical Research Library and Archives, New York; Dingwall, "Report of a Series," 100–1.
24 Dingwall, "Report of a Series," 101.
25 "Sitting at 10 Lime Street, Fifth Séance."
26 Ibid.
27 "Sitting at 10 Lime Street," January 7, 1925, Margery Fonds, box 119, file 4, American Society for Psychical Research Library and Archives, New York; Dingwall, "Report of a Series," 107.
28 Jane Bennett, *Vibrant Matter: A Political Ecology of Things* (Durham, NC: Duke University Press, 2010), 20.
29 This flurry of activity is noted in T.G. Hamilton, "Margery in Winnipeg: Three Séances of December 1926," in *Proceedings of the American Society for Psychical Research: The Margery Mediumship,* vol. 2 (New York: American Society for Psychical Research, 1933), 556. It is also summarized more cynically in Walter Franklin Prince, "Some Highlights," *Boston Society for Psychical Research Bulletin: The "Walter"-"Kerwin" Thumb Prints,* 22 (April 1934): 79–85.
30 Prince, "Some Highlights," 79–85.
31 J. Malcolm Bird, "Walter's 'Talking Head,'" in *Proceedings of the American Society for Psychical Research: The Margery Mediumship,* vol. 1 (New York: American Society for Psychical Research, 1928), 16.
32 "Sitting at 10 Lime Street," February 24, 1925; "Sitting at 10 Lime Street," February 26, 1925, Margery Fonds, box 119, file 5, American Society for Psychical Research Archives and Special Collections, New York.
33 "Sitting at 10 Lime Street," February 24, 1925.
34 "Sitting at 10 Lime Street," February 27, 1925, Margery Fonds, box 119, file 5, American Society for Psychical Research Archives and Special Collections, New York.
35 "Sitting at 10 Lime Street," March 2, 1925, Margery Fonds, box 119, file 5, American Society for Psychical Research Archives and Special Collections, New York.
36 Bird, "Walter's 'Talking Head,'" 21–22.
37 J. Malcolm Bird, "The Climax of the A.S.P.R. Sittings," in *Proceedings of the American Society for Psychical Research: The Margery Mediumship,* vol. 1 (New York: American Society for Psychical Research, 1928), 214–15.

38 Ibid., 218, 223.

39 "Sitting at 10 Lime Street," August 13, 1925, Margery Fonds, box 119, file 5, American Society for Psychical Research Archives and Special Collections, New York.

40 Ibid.; Bird, "Climax of the A.S.P.R. Sittings," 224.

41 Hamilton, "General Summary of Poole's Development."

42 Ibid.

43 T.G. Hamilton, "Experimentation for Materializations with Mrs. Elizabeth Poole, 1923, 1924: Wax Moulds," Hamilton Family Fonds, MSS14 (A.79-41), box 9, folder 14, University of Manitoba Archives and Special Collections, Winnipeg.

44 Ibid.

45 Ibid.

46 Ibid.

47 T.G. Hamilton, "Teleplasmic Phenomenon in Winnipeg," *Psychic Science: Quarterly Transactions of the British College* 8, 3 (1929): 183.

48 T.G. Hamilton, "Divisions of the Mary M. Teleplasms (Followed Chronologically)," Hamilton Family Fonds, MSS14 (A.79-41), box 15, folder 10, University of Manitoba Archives and Special Collections, Winnipeg.

49 Hamilton, "Teleplasmic Phenomenon in Winnipeg," 201, 205.

50 Hamilton, "Divisions of the Mary M. Teleplasms."

51 Ibid.

52 T.G. Hamilton, "Research Shows Many Functions for Teleplasm," Hamilton Family Fonds, MSS14 (A.79-41), box 15, folder 6, University of Manitoba Archives and Special Collections, Winnipeg.

53 Harry Green as Dr. James Hyslop, quoted in Bruce Chown, "Report of Conversation between Dr. James Hyslop through Ewan and Dr. T.G. Hamilton," February 10, 1932, Hamilton Family Fonds, MSS14 (A.79-41), box 16, folder 8, University of Manitoba Archives and Special Collections, Winnipeg.

54 Ibid.

55 T.G. Hamilton, "The Second Lucy Series" (August 1931 to November 27, 1931), Hamilton Family Fonds, MSS14 (A.79-41), box 16, folder 7, University of Manitoba Archives and Special Collections, Winnipeg.

56 Hamilton, "Divisions of the Mary M. Teleplasms," 11; "More Instructions," Seance notes, February 17, 1929, Hamilton Family Fonds, MSS14 (A.79-41), box 15, folder 14, University of Manitoba Archives and Special Collections, Winnipeg.

57 "Walter Foretells Two Things," Seance notes, March 19, 1929, Hamilton Family Fonds, MSS14 (A.79-41), box 15, folder 16, University of Manitoba Archives and Special Collections, Winnipeg.

58 Seance notes, July 10, 1929, Hamilton Family Fonds, MSS14 (A.79-41), box 15, folder 16, University of Manitoba Archives and Special Collections, Winnipeg.

59 "Serious Words from Black Hawk," Seance notes, July 17, 1929, Hamilton Family Fonds, MSS14 (A.79-41), box 15, folder 16, University of Manitoba Archives and Special Collections, Winnipeg.

60 Hamilton, "Article 3."

61 Ibid.

62 Latour, *Reassembling the Social*, 71–72.

63 Hamilton, "Research Shows Many Functions."

64 Ibid.

65 Donna J. Haraway, *Modest_Witness@Second_Millennium.FemaleMan©_Meets_OncoMouse™: Feminism and Technoscience* (New York: Routledge, 1997), 32.
66 T.G. Hamilton, "E.M. Trance and Visions," February 1, 1933, Hamilton Family Fonds, MSS14 (A.79-41), box 16, folder 11, University of Manitoba Archives and Special Collections, Winnipeg.
67 T.G. Hamilton, "The Mary M. Teleplasm of October 27th, 1929," *Psychic Science: Quarterly Transactions of the British College* 10, 4 (1932): 255.
68 Ibid.
69 Sofie Lachapelle, *Investigating the Supernatural: From Spiritism and Occultism to Psychical Research and Metaphysics in France, 1853-1931* (Baltimore, MD: John Hopkins University Press, 2011), 28, 117.
70 T.G. Hamilton, "E.M.'s Attitude toward Her Trance Personalities," Hamilton Family Fonds, MSS14 (A.79-41), box 9, folder 9, University of Manitoba Archives and Special Collections, Winnipeg.
71 T.G. Hamilton, "The Handwriting on the Wall," June 10, 1931, Hamilton Family Fonds, MSS14 (A.79-41), box 16, folder 6, University of Manitoba Archives and Special Collections, Winnipeg.
72 Roger Luckhurst, *The Invention of Telepathy, 1870-1901* (Oxford: Oxford University Press, 2002), 26–28; Jeffrey Sconce, *Haunted Media: Electronic Presence from Telegraphy to Television* (Durham, NC: Duke University Press, 2000), 60.
73 T.G. Hamilton, "IX Records: Katie King," November 12, 1930, Hamilton Family Fonds, MSS14 (A.79-41), box 16, folder 4, University of Manitoba Archives and Special Collections, Winnipeg.
74 T.G. Hamilton, "Katie King Manifestations in the Mary M. Experiments," *Psychic Science: Quarterly Transactions of the British College* 11, 4 (1933): 276.
75 T.G. Hamilton, "Group IV: The Shining Garment," January 25, 1931, Hamilton Family Fonds, MSS14 (A.79-41), box 16, folder 5, University of Manitoba Archives and Special Collections, Winnipeg.
76 "Sitting #358," Seance notes, February 5, 1933, Hamilton Family Fonds, MSS14 (A.79-41), box 16, folder 11, University of Manitoba Archives and Special Collections, Winnipeg.
77 Hamilton, "IX Records"; Hamilton, "E.M. Trance and Visions"; T.G. Hamilton, "The Shell of Katie Appears, Sitting #329," April 23, 1933, Hamilton Family Fonds, MSS14 (A.79-41), box 16, folder 11, University of Manitoba Archives and Special Collections, Winnipeg.
78 Lorraine Daston and Peter Galison, *Objectivity* (New York: Zone Books, 2007), 37.
79 Dingwall, "Report of a Series," 98; Hamilton, "Margery in Winnipeg," 560–61; Margaret L. Hamilton, "Outline of the Sequence of the Hamilton Experiments and Investigation of Psychic Phenomena," Hamilton Family Fonds, MSS14 (A.79-41), box 15, folder 7, University of Manitoba Archives and Special Collections, Winnipeg.
80 T.G. Hamilton, "The C.H. Spurgeon Case: Evidence Pointing to Survival and Continued Activity," *Light: A Journal of Psychical, Occult and Mystical Research* 53, 2752 (1933): 628.
81 T.G. Hamilton, "March 15th, 1928: The Young Man Asks for an Electric Bell," Hamilton Family Fonds, MSS14 (A.79-41), box 15, folder 11, University of Manitoba Archives and Special Collections, Winnipeg.
82 Seance notes, October 27, 1929, Hamilton Family Fonds, MSS14 (A.79-41), box 15, folder 10, University of Manitoba Archives and Special Collections, Winnipeg.
83 Ibid.

84 E.E. Dudley, "Psychics versus Mediums: The Extremes of the Psychical Spectrum as Seen by an Observer of Many Years' Experience," *Psychic Research* 23, 2 (1929): 68–69.

85 Bird, *Margery the Medium,* 264.

86 Ibid., 350.

87 Agreement between Eric J. Dingwall and Dr. L.R.G. Crandon, December 30, 1924, Margery Fonds, box 119, file 4, American Society for Psychical Research Library and Archives, New York.

88 Dudley, "Psychics versus Mediums," 69–70.

89 T.G. Hamilton, "A Costly Mistake," October 20, 1929, Hamilton Family Fonds, MSS14 (A.79-41), box 16, folder 1, University of Manitoba Archives and Special Collections, Winnipeg.

90 Ibid.

91 Ibid. For more on the dynamic relationship between spirits in the Hamilton seance room, see Beth A. Robertson, "Feminine Apparitions and Other Ghostly Teleplasm: Contesting and Constructing Womanliness in the Séances of Dr. T.G. Hamilton, 1918-1935," in *Spirit, Faith and Church: Women's Experiences in the English-Speaking World, 17th-21st Centuries,* ed. Laurence Lux-Sterritt and Claire Sorin (Newcastle-upon-Tyne, UK: Cambridge Scholars, 2012), 153–54.

92 Haraway, *Modest_Witness,* 31.

93 Hamilton, "General Summary of Poole's Development," 2–3; Hamilton, "Article III," 4.

94 Ann Taves, *Fits, Trances and Visions: Experiencing Religion and Explaining Experience from Wesley to James* (Princeton, NJ: Princeton University Press, 1999), 176; John B. Buescher, *The Other Side of Salvation: Spiritualism and Nineteenth-Century Religious Experience* (Boston: Skinner House Books, 2004), 85.

95 T.G. Hamilton, "A Study of the Winnipeg Group Mediumship in Its Relation to the Dawn Teleplasms," *Psychic Research* 29, 2 (1934): 120.

96 Hamilton, "E.M. Trance and Visions."

97 "Second Ectoplasmic Structure," Seance notes, August 15, 1928, Hamilton Family Fonds, MSS14 (A.79-41), box 15, folder 12, University of Manitoba Archives and Special Collections, Winnipeg.

98 Sconce, *Haunted Media,* 44.

99 Fidella Dinsmore Papa to Jenny O'Hara Pincock, June 4, 1936, Maines Pincock Family Fonds, GA64, series 1.1, file 9, University of Waterloo Library, Waterloo.

100 Seance notes, October 17, 1929, Maines Pincock Family Fonds, GA64, series 4, file 93, University of Waterloo Library, Waterloo; O'Hara Pincock, *Trails of Truth,* 16, 22.

101 O'Hara Pincock, *Trails of Truth,* 22, 53.

102 Seance notes, October 17, 1929.

103 Seance notes, January 26, 1930, Manes Pincock Family Fonds, GA64, series 4, file 94, University of Waterloo Library, Waterloo.

104 T.G. Hamilton, quoted in L. Hamilton, "Comments on the Life and Interests of TGH," Hamilton Family Fonds, MSS14 (A.79-41), box 1, folder 4, University of Manitoba Archives and Special Collections, Winnipeg.

105 Hamilton, "General Summary of Poole's Development."

106 "More Instructions and Preparation," Seance notes, March 27, 1929, Hamilton Family Fonds, MSS14 (A.79-41), box 15, folder 14, University of Manitoba Archives and Special Collections, Winnipeg.

107 "A Flash by Chance: Black and White Teleplasm," Seance notes, April 7, 1929, Hamilton Family Fonds, MSS14 (A.79-41), box 15, folder 14, University of Manitoba Archives and Special Collections, Winnipeg.

108 Judith Butler, *Bodies That Matter: On The Discursive Limits of "Sex"* (New York: Routledge, 1993), 53.

109 Jill Galvan, "The Victorian Post-human: Transmission, Information and the Séance," in *The Ashgate Research Companion to Nineteenth-Century Spiritualism and the Occult*, ed. Tatiana Kontou and Sarah Willburn (Surrey, UK: Ashgate, 2012), 80.

110 Sconce, *Haunted Media*, 27.

111 T.G. Hamilton, "The Mediumship of Mary Marshall," Hamilton Family Fonds, MSS14 (A.79-41), box 15, folder 12, University of Manitoba Archives and Special Collections, Winnipeg.

112 Seance notes, February 15, 1927, Margery Fonds, box 119, file 8, American Society for Psychical Research Library and Archives, New York.

113 Bird, "Climax of the A.S.P.R. Sittings," 219.

114 Hamilton, "Article 3."

115 J. Malcolm Bird, "The A.S.P.R. Project: Séances of June, 1925," in *Proceedings of the American Society for Psychical Research: The Margery Mediumship*, vol. 1 (New York: American Society for Psychical Research, 1928), 146.

116 "Sitting at 10 Lime Street," March 6, 1926, Margery Fonds, box 119, file 7, American Society for Psychical Research Library and Archives, New York.

117 Hamilton, "Mary M. Teleplasm," 249.

118 Bird, "Climax of the A.S.P.R. Sittings," 232.

119 Dingwall's books on sexology include *The Girdle of Chastity: A Medico-Historical Study* (Paris: Le Divan, 1923), *Male Infibulation* (London: John Bale and Sons and Danielsson, 1925), and *Some Human Oddities: Studies in the Queer, the Uncanny and the Fanatical* (London: Home and Van Thal, 1947).

120 Anne Fausto-Sterling, *Sexing the Body: Gender Politics and the Construction of Sexuality* (New York: Basic Books, 2000), 177–78; Sarah S. Richardson, *Sex Itself: The Search for Male and Female in the Human Genome* (Chicago: University of Chicago Press, 2013), 63; Chandak Sengoopta, *The Most Secret Quintessence of Life: Sex, Glands and Hormones, 1850-1950* (Chicago: University of Chicago Press, 2006) 69–204.

121 Dingwall, "Report of a Series," 108.

122 Daniel Day Walton, "The McComas-Wood-Dunlap Commission and Its Report," in *Proceedings of the American Society for Psychical Research: The Margery Mediumship*, vol. 2 (New York: American Society for Psychical Research, 1933), 715.

123 J. Malcolm Bird, "Another Committee," in *Proceedings of the American Society for Psychical Research: The Margery Mediumship*, vol. 1 (New York: American Society for Psychical Research, 1928), 369; "Sitting at 10 Lime Street," January 26, 1926, Margery Fonds, box 119, file 7, American Society for Psychical Research Library and Archives, New York.

124 "Sitting at 10 Lime Street," January 27, 1926, Margery Fonds, box 119, file 7, American Society for Psychical Research Library and Archives, New York.

125 "Sitting at 10 Lime Street," January 28, 1926, Margery Fonds, box 119, file 7, American Society for Psychical Research Library and Archives, New York.

126 "Sitting at 10 Lime Street," January 30, 1926, Margery Fonds, box 119, file 7, American Society for Psychical Research Library and Archives, New York.

127 Bird, "Another Committee," 371.

128 Ibid., 383–84.

129 Walton, "McComas-Wood-Dunlap Commission," 737.

130 Ibid.

131 "Sitting at 10 Lime Street," January 30, 1926.

132 Dingwall, "Report of a Series," 104–5.
133 Butler, *Bodies That Matter,* 60–61.
134 Ibid. 63, 85.
135 Ibid., 89.
136 Ibid.
137 Haraway, *Modest_Witness,* 24.
138 Donna Haraway, *Simians, Cyborgs and Women: The Reinvention of Nature* (New York: Routledge, 1991), 149, 177, 195.

The Knot Unravelled

1 Sigmund Freud, "The Uncanny" (1919), in *An Infantile Neurosis and Other Works: The Standard Edition of the Complete Psychoanalytic Works of Sigmund Freud,* vol. 17, *1917–19,* trans. James Strachey (London: Hogarth, 1955), 240.
2 "Psychic Field of Research Loses Leader: Dr. T. Glen Hamilton, Noted Investigator and Former M.L.A., Victim of Heart Ailment," undated newspaper clipping, Maines Pincock Family Fonds, GA64, series 6, file 169, University of Waterloo Library, Waterloo; "Obituaries: Dr. T. Glendenning Hamilton," *Canadian Medical Association Journal* (June 1935), Hamilton Family Fonds, MSS14 (79–41), box 2, folder 6, University of Manitoba Archives and Special Collections, Winnipeg; Elwood Worcester, "Dr. Prince as a Psychic Researcher and as a Psychiatrist," in *Walter Franklin Prince: A Tribute to His Memory,* ed. Boston Society for Psychical Research (Boston: Boston Society for Psychical Research, 1935), 9–10.
3 A.S.P.R. Research Committee of the Board of Trustees, "Preface," in *Proceedings of the American Society for Psychical Research: The Margery Mediumship,* vol. 2 (New York: American Society for Psychical Research, 1933), i-ii.
4 Henry Clay McComas, *Ghosts I Have Talked With* (Baltimore, MD: Williams and Wilkins Company, 1935).
5 Jenny O'Hara Pincock to Henry Remmers, April 5, 1944, Maines Pincock Family Fonds, GA64, series 1.1, file 17, University of Waterloo Library, Waterloo.
6 "Famed Medium, Ontario-Born 'Margery,' Dies," *St. Catharines Standard,* November 3, 1941, Maines Pincock Family Fonds, GA64, series 6, file 169, University of Waterloo Library, Waterloo; Thomas R. Tietze, *Margery* (New York: Harper and Row, 1973), 183–84; John F. Kasson, *Houdini, Tarzan and the Perfect Man: The White Male Body and the Challenge of Modernity in America* (New York: Hill and Wang, 2001), 153.
7 David Howes, "Introduction: The Revolving Sensorium," in *The Sixth Sense Reader,* ed. David Howes (Oxford: Berg, 2009), 6–8; H.M. Collins and T.J. Pinch, "The Construction of the Paranormal: Nothing Unscientific Is Happening," *On the Margins of Science: The Social Construction of Rejected Knowledge,* ed. Roy Wallis (Keele, UK: University of Keele Press, 1979), 242–44; R. Laurence Moore, *In Search of White Crows: Spiritualism, Parapsychology and American Culture* (New York: Oxford University Press, 1977), 185–220. First-hand accounts of their establishment of the lab at Duke include J.B. Rhine, *New Frontiers of the Mind: The Story of the Duke Experiments* (New York: Farrar and Rinehart, 1937); and Louisa E. Rhine, *Mind over Matter* (New York: Macmillan, 1970), 70–100.
8 Jenny O'Hara Pincock to Henry Remmers, April 5, 1944, Maines Pincock Family Fonds, GA64, series 1.1, file 17, University of Waterloo Library, Waterloo.
9 Hereward Carrington, *The Invisible World* (New York: Beechhurst Press and B. Ackerman, 1946), 148.

10 Margaret L. Hamilton, "Notes on Mr. Harry A.V. Green, Co-worker with Dr. T. Glen Hamilton," Hamilton Family Fonds, MSS14 (A.79-41), box 1, folder 1, University of Manitoba Archives and Special Collections, Winnipeg.
11 "The TGH Teleplasm," Seance notes, May 22, 1939, Hamilton Family Fonds, MSS14 (A.79-41), box 15, folder 10, University of Manitoba Archives and Special Collections, Winnipeg.
12 "Introduction to the TGH-Dawn Scripts (August 1943-April 1944)," Hamilton Family Fonds, MSS14 (A.79-41), box 15, folder 6, University of Manitoba Archives and Special Collections, Winnipeg.
13 E.J. Dingwall, "The Need for Responsibility in Parapsychology: My Sixty Years in Psychical Research" (1971), in *A Skeptic's Handbook of Parapsychology*, ed. Paul Kurtz (New York: Prometheus Books, 1985), 161.
14 Ibid., 162, 171.
15 Ibid., 168.
16 Ibid., 162–63.
17 Donna J. Haraway, *Modest_Witness@Second_Millennium.FemaleMan©_Meets_OncoMouse™: Feminism and Technoscience* (New York: Routledge, 1997), 24–25.

Bibliography

Archival Collections

American Society for Psychical Research Library and Archives, New York. Margery Fonds.

Cambridge University Library, London, England. Society for Psychical Research Fonds, MS39/14.

University of Manitoba Archives and Special Collections, Winnipeg. Hamilton Family Fonds, MSS14 (A.79-41) and PC12 (A.79-41).

University of Waterloo Library, Waterloo. Maines Pincock Family Fonds, GA64.

Other Sources

Albanese, Catherine L. *A Republic of Mind and Spirit: A Cultural History of American Metaphysical Religion*. New Haven, CT: Yale University Press, 2007.

Allen, Glen Scott. *Master Mechanics and Wicked Wizards: Images of the American Scientist as Hero and Villain from Colonial Times to the Present*. Amherst: University of Massachusetts Press, 2009.

Appignanesi, Lisa. *Mad, Bad and Sad: A History of Women and the Mind Doctors*. 2007. Reprint, New York: W.W. Norton and Company, 2008.

Arondekar, Anjali. "Without a Trace: Sexuality and the Colonial Archive." *Journal of the History of Sexuality* 14, 1 (2005): 10–27. http://dx.doi.org/10.1353/sex.2006.0001.

A.S.P.R. Research Committee of the Board of Trustees. "Preface." In *Proceedings of the American Society for Psychical Research: The Margery Mediumship*, vol. 2, 491–92. New York: American Society for Psychical Research, 1933.

Austin, B.F. "Foreword." In Jenny O'Hara Pincock, *Trails of Truth*, 3–4. Los Angeles, CA: Austin, 1930.

Banks, Rosemarie K. "Staging the 'Native': Making History in American Theatre Culture, 1828–1932." *Theatre Journal* 15, 4 (1993): 461–86.

Basham, Diana. *The Trial of Woman: Feminism and the Occult Sciences in Victorian Literature and Society*. Bastingstoke, UK: Macmillan, 1992. http://dx.doi.org/10.1057/9780230374010.

Bederman, Gail. *Manliness and Civilization: A Cultural History of Gender and Race in the United States, 1880–1917*. Chicago: University of Chicago Press, 1995. http://dx.doi.org/10.7208/chicago/9780226041490.001.0001.

Benjamin, Marina. "Introduction." In *Science and Sensibility: Gender and Scientific Enquiry, 1780–1945*, ed. Marina Benjamin, 1–24. Oxford: Blackwell, 1991.

Bennett, Jane. *Vibrant Matter: A Political Ecology of Things*. Durham, NC: Duke University Press, 2010.

Benson, John. *The Working Class in Britain, 1850–1939*. London: I.B. Taurus, 2003.

Bhabha, Homi K. *The Location of Culture*. London: Routledge, 1994.

Bird, J. Malcolm. "Another Committee." In *Proceedings of the American Society for Psychical Research: The Margery Mediumship*, vol. 1, 365–89. New York: American Society for Psychical Research, 1928.

–. "The A.S.P.R. Project: Séances of June, 1925." In *Proceedings of the American Society for Psychical Research: The Margery Mediumship*, vol. 1, 143–70. New York: American Society for Psychical Research, 1928.

–. "A.S.P.R. Seances: Second Series." In *Proceedings of the American Society for Psychical Research: The Margery Mediumship*, vol. 1, 171–87. New York: American Society for Psychical Research, 1928.

–. "The Bell-Boxes of Early 1925." In *Proceedings of the American Society for Psychical Research: The Margery Mediumship*, vol. 1, 114–22. New York: American Society for Psychical Research, 1928.

–. "The Climax of the A.S.P.R. Sittings." In *Proceedings of the American Society for Psychical Research: The Margery Mediumship*, vol. 1, 214–41. New York: American Society for Psychical Research, 1928.

–. "The Code-Hoagland Episode." In *Proceedings of the American Society for Psychical Research: The Margery Mediumship*, vol. 1, 85–113. New York: American Society for Psychical Research, 1928.

–. "The Curious History of the 'Hole Bell-Box.'" In *Proceedings of the American Society for Psychical Research: The Margery Mediumship*, vol. 1, 123–31. New York: American Society for Psychical Research, 1928.

–. "Dr. Richardson's Voice-Control Machine." In *Proceedings of the American Society for Psychical Research: The Margery Mediumship*, vol. 1, 258–76. New York: American Society for Psychical Research, 1928.

–. "The Glass Cabinet and the Wired Control." In *Proceedings of the American Society for Psychical Research: The Margery Mediumship*, vol. 1, 312–23. New York: American Society for Psychical Research, 1928.

–. "Informal Tests of the Walter Voice." In *Proceedings of the American Society for Psychical Research: The Margery Mediumship*, vol. 1, 24–30. New York: American Society for Psychical Research, 1928.

–. *Margery the Medium*. Boston: Small, Maynard and Company, 1925.

–. "The Margery Mediumship: An Experiment in Fraud-Proof Control of a New Type." *Journal of the American Society for Psychical Research* 20, 11 (1926): 674–82.

–. "Paraffin Gloves under the Co-mediumship of Dr. Hardwicke." In *Proceedings of the American Society for Psychical Research: The Margery Mediumship*, vol. 1, 447–64. New York: American Society for Psychical Research, 1928.

–. "The Rehabilitation of the Scales." In *Proceedings of the American Society for Psychical Research: The Margery Mediumship,* vol. 1, 417–27. New York: American Society for Psychical Research, 1928.

–. "The Teleplasmic Hand." In *Proceedings of the American Society for Psychical Research: The Margery Mediumship,* vol. 1. New York: American Society for Psychical Research, 1928.

–. "Teleplasmic Thumbprints – V: Further Particulars of the Experiments in and out of the Séance Room." *Psychic Research* 22, 10 (1928): 563–71.

–. "Walter's 'Talking Head.'" In *Proceedings of the American Society for Psychical Research: The Margery Mediumship,* vol. 1, 16–23. New York: American Society for Psychical Research, 1928.

Bird, S. Elizabeth. "Introduction: Constructing the Indian, 1830s-1990s." In *Dressing in Feathers: The Construction of the Indian in American Popular Culture,* 1–12. Boulder, CO: Westview, 1996.

Blum, Deborah. *Ghost Hunters: William James and the Search for Scientific Proof of Life after Death.* New York: Penguin, 2006.

Bourke, Joanna. *Dismembering the Male: Men's Bodies, Britain and the Great War.* Chicago: University of Chicago Press, 1996.

–. *Working-Class Cultures in Britain, 1890–1960: Gender, Class and Ethnicity.* London: Routledge, 1994.

Bowler, Peter J. *Science for All: The Popularization of Science in Early Twentieth-Century Britain.* Chicago: University of Chicago Press, 2009. http://dx.doi.org/10.7208/chicago/9780226068664.001.0001.

Brain, Robert Michael. "Materializing the Medium: Ectoplasm and the Quest for Supranormal Biology in Fin-de-Siècle Science and Art." In *Vibratory Modernism,* ed. Anthony Enns and Shelley Trower, 115–44. Houndsmills, UK: Palgrave Macmillan, 2013.

Brandon, Ruth. *The Spiritualists: The Passion for the Occult in the Nineteenth and Twentieth Centuries.* Buffalo: Prometheus Books, 1984.

Braude, Ann. *Radical Spirits: Spiritualism and Women's Rights in Nineteenth-Century America.* 2nd ed. Bloomington: Indiana University Press, 2001.

Bright, David. *The Limits of Labour: Class Formation and the Labour Movement in Calgary, 1883–1929.* Vancouver: UBC Press, 1998.

Brooks, Daphne A. *Bodies in Dissent: Spectrality Performances of Race and Freedom, 1850–1910.* Durham, NC: Duke University Press, 2006.

Brown, Callum. *Religion and Society in Twentieth-Century Britain.* Harlow, UK: Pearson Education, 2006.

Brown, Hunter. *William James on Radical Empiricism and Religion.* Toronto: University of Toronto Press, 2000.

Buescher, John B. *The Other Side of Salvation: Spiritualism and Nineteenth-Century Religious Experience.* Boston: Skinner House Books, 2004.

Burton, Antoinette. *Dwelling in the Archive: Women Writing House, Home and History in Late Colonial India.* Oxford: Oxford University Press, 2003. http://dx.doi.org/10.1093/acprof:oso/9780195144253.001.0001.

Butler, Judith. *Bodies That Matter: On the Discursive Limits of "Sex."* New York: Routledge, 1993.

Button, William H. "The Margery Mediumship: A Solus Sitting for Thumb Print March 11th, 1931 under Additional Techniques of Control." In *The Margery Mediumship:*

Records of Supernormal Production of Thumbprints, ed. R.J. Tillyard, 13–17. New York: American Society for Psychical Research, 1931.

Carrington, Hereward. *The Invisible World.* New York: Beechhurst Press and B. Ackerman, 1946.

Charet, F.X. *Spiritualism and the Foundations of C.G. Jung's Psychology.* New York: State University of New York Press, 1993.

Classen, Constance. *The Color of Angels: Cosmology, Gender and the Aesthetic Imagination.* London: Routledge, 1998.

–. *The Deepest Sense: A Cultural History of Touch.* Urbana: University of Illinois Press, 2012.

–. "McLuhan in the Rainforest: The Sensory Worlds of Oral Cultures." In *Empire of the Senses: The Sensual Culture Reader,* ed. David Howes, 147–63. Oxford: Berg, 2005.

Classen, Constance, David Howe, and Anthony Synnott. *Aroma: The Cultural History of Smell.* London: Routledge, 1994.

Clavin, Patricia. *Securing the World Economy: The Reinvention of the League of Nations, 1920–1946.* Oxford: Oxford University Press, 2013. http://dx.doi.org/10.1093/acprof:oso/9780199577934.001.0001.

Code, Lorraine. *What Can She Know? Feminist Theory and the Construction of Knowledge.* Ithaca, NY: Cornell University Press, 1991.

Collins, H.M., and T.J. Pinch. "The Construction of the Paranormal: Nothing Unscientific Is Happening." In *On the Margins of Science: The Social Construction of Rejected Knowledge,* ed. Roy Wallis, 237–70. Keele, UK: University of Keele, 1979. http://dx.doi.org/10.1111/j.1467-954X.1979.tb00064.x.

Comacchio, Cynthia. *The Infinite Bonds of Family: Domesticity in Canada, 1850–1940.* Toronto: University of Toronto Press, 1999.

Connor, Steven. "The Machine in the Ghost: Spiritualism, Technology and Direct Voice." In *Ghosts: Deconstruction, Psychoanalysis, History,* ed. Peter Buse and Andrew Stott, 203–25. London: Macmillan, 1999.

Cook, Ramsay. *The Regenerators: Social Criticism in Late Victorian English Canada.* Toronto: University of Toronto Press, 1985.

Crandon, Le Roi. "The Margery Mediumship: Supernormal Cognition, Subjective and Objective." *Journal of the American Society for Psychical Research* 20, 6 (1926): 321–33.

Crary, Jonathan. *Techniques of the Observer: On Vision and Modernity in the Nineteenth Century.* 1990. Reprint, Cambridge, MA: MIT Press, 1995.

Curthoys, Ann, and Marilyn Lake. "Introduction." In *Connected Worlds: History in Transnational Perspective,* ed. Ann Curthoys and Marilyn Lake, 5–20. Canberra: Australian National University E Press, 2005.

–, eds. *Connected Worlds: History in Transnational Perspective.* Canberra: Australian National University E Press, 2005.

Curtis, Heather D. *Faith in the Great Physician: Suffering and Divine Healing in American Culture.* Baltimore, MD: John Hopkins University Press, 2007.

Daston, Lorraine, and Peter Galison. *Objectivity.* New York: Zone Books, 2007.

Dean, Joanna. *Religious Experience and the New Woman: The Life of Lily Dougall.* Bloomington: Indiana University Press, 2007.

DeBrath, Simon. "Chinese Script and Voices." *Psychic Science: Quarterly Transactions of the British College* 8, 3 (1929): 213–17.

–. "Editorial Notes." *Psychic Science: Quarterly Transactions of the British College* 9, 4 (1931): 233–42.

–. "Editorial Notes." *Psychic Science: Quarterly Transactions of the British College* 12, 4 (1934): 237–43.

Deloria, Philip J. *Indians in Unexpected Places*. Lawrence: University Press of Kansas, 2004.

–. *Playing Indian*. New Haven, CT: Yale University Press, 1998.

del Pilar Blanco, María, and Esther Peeren. "Spectropolitics: Ghosts of the Global Contemporary." In *The Spectralities Reader: Ghosts and Haunting in Contemporary Cultural Theory*, ed. María del Pilar Blanco and Esther Peeren, 91-102. London: Bloomsbury, 2013.

Deutsch, Sarah. *Women and the City: Gender, Space and Power in Boston, 1870-1940*. Oxford: Oxford University Press, 2002.

Dixon, Joy. *Divine Feminine: Theosophy and Feminism in England*. Baltimore, MD: John Hopkins University Press, 2001.

Dolby, R.G.A. "Reflections on Deviant Science." In *On the Margins of Science: The Social Construction of Rejected Knowledge*, ed. Roy Wallis, 9–47. Keele, UK: University of Keele, 1979.

Douglas, Ann. *Terrible Honesty: Mongrel Manhattan in the 1920s*. New York: Noonday, 1995.

Doyle, Arthur Conan. *The History of Spiritualism*. Vol. 1. London: Cassel and Company, 1926.

Dingwall, E.J. *The Girdle of Chastity: A Medico-Historical Study*. Paris: Le Divan, 1923.

–. *How to Go to a Medium: A Manual of Instruction*. London: Kegan Paul, Trench, Trubner and Company, 1922.

–. *Male Infibulation*. London: John Bale and Sons and Danielsson, 1925.

–. "The Need for Responsibility in Parapsychology: My Sixty Years in Psychical Research" (1971). In *A Skeptic's Handbook of Parapsychology*, ed. Paul Kurtz, 161–74. New York: Prometheus Books, 1985.

–. "A Report of a Series of Sittings with the Medium Margery, June 1926." In *Proceedings of the Society for Psychical Research*, vol. 36, 79–155. Glasgow, UK: Robert MacLehose and Company, 1928.

–. *Some Human Oddities: Studies in the Queer, the Uncanny and the Fanatical*. London: Home and Van Thal, 1947.

Driesch, Hans. "Walter Prince as a Scientist." In *Walter Franklin Prince: A Tribute to His Memory*, ed. Boston Society for Psychical Research, 34–39. Boston: Boston Society for Psychical Research, 1935.

Dudley, E.E. "The Butler Apparatus." In *Proceedings of the American Society for Psychical Research: The Margery Mediumship*, vol. 2, 588–606. New York: American Society for Psychical Research, 1933.

–. "Five Outstanding Records of 1927." In *Proceedings of the American Society for Psychical Research: The Margery Mediumship*, vol. 2, 607-25. New York: American Society for Psychical Research, 1933.

–. "Miscellaneous Episodes of Late 1926." In *Proceedings of the American Society for Psychical Research: The Margery Mediumship*, vol. 2, 540–55. New York: American Society for Psychical Research, 1933.

–. "Photographic Experiments in 1927-8." In *The Margery Mediumship: Proceedings of the American Society for Psychical Research*, vol. 2, 699–714. New York: American Society for Psychical Research, 1933.

–. "Post-Trance Recording of Séance Cognitions." In *Proceedings of the American Society for Psychical Research: The Margery Mediumship*, vol. 2, 749–62. New York: American Society for Psychical Research, 1933.

–. "Psychics versus Mediums: The Extremes of the Psychical Spectrum as Seen by an Observer of Many Years' Experience." *Psychic Research* 23, 2 (1929): 65–75.

–. "Telekinesis in 1928 and 1929." In *Proceedings of the American Society for Psychical Research: The Margery Mediumship*, vol. 2, 672–98. New York: American Society for Psychical Research, 1933.

–. "Telekinesis of Early 1927." In *Proceedings of the American Society for Psychical Research: The Margery Mediumship*, vol. 2, 568–87. New York: American Society for Psychical Research, 1933.

–. "Telekinesis of Later 1927." In *Proceedings of the American Society for Psychical Research: The Margery Mediumship*, vol. 2, 653–71. New York: American Society for Psychical Research, 1933.

–. "Teleplasmic Structures and Telekinesis." In *Proceedings of the American Society for Psychical Research: The Margery Mediumship*, vol. 2, 626–52. New York: American Society for Psychical Research, 1933.

Duffin, Jacalyn. *History of Medicine: A Scandalously Short Introduction*. 2nd ed. Toronto: University of Toronto Press, 2010.

Dyck, Noel. "Canadian Anthropology and the Ethnography of 'Indian' Administration." In *Historicizing Canadian Anthropology*, ed. Julia Harrison and Regina Darnell, 78–106. Vancouver: UBC Press, 2006.

Erickson, Mark. *Science, Culture and Society: Understanding Science in the Twenty-First Century*. Cambridge, UK: Polity, 2005.

Faue, Elizabeth. "Re-imagining Labour: Gender and New Directions in Labour and Working-Class History." In *Re-thinking U.S. Labour History: Essays on the Working-Class Experience, 1756–2009*, ed. Donna T. Haverty-Stacke and Daniel J. Walkowitz, 266–88. New York: Continuum International Publishing Group, 2010.

Fausto-Sterling, Anne. *Sexing the Body: Gender Politics and the Construction of Sexuality*. New York: Basic Books, 2000.

Febvre, Lucien. "Comment reconstituer la vie affective d'autrefois? La sensibilité et l'histoire." *Annales d'histoire sociale* 3, 1–2 (1941): 5–20.

Foucault, Michel. *The Birth of the Clinic: An Archaeology of Medical Perception*. Trans. A.M. Sheridan. 1973; Reprint, London: Routledge, 2008.

–. *Discipline and Punish: The Birth of the Prison*. Trans. Alan Sheridan. New York: Vintage Books, 1977.

–. "Nietzsche, Genealogy, History." In *Aesthetics, Method and Epistemology*, ed. James D. Faubion, 369–91. New York: New Press, 1998.

Fox Keller, Evelyn. *Reflections on Gender and Science*. New Haven, CT: Yale University Press, 1985.

–. *Secrets of Life, Secrets of Death: Essays on Language, Gender and Science*. New York: Routledge, 1992.

Freud, Sigmund. "The Uncanny" (1919). In *An Infantile Neurosis and Other Works: The Standard Edition of the Complete Psychoanalytic Works of Sigmund Freud*, vol. 17, *1917–19*, trans. James Strachey, 217–56. London: Hogarth, 1955.

Gaertner, Irving R. "The Acting Power of Light as a Disturbing Element in Physical Mediumship." *Psychic Research* 25, 1 (1931): 3–4.

Galvan, Jill. *The Sympathetic Medium: Feminine Channelling, the Occult and Communication Technologies, 1859–1919*. Ithaca, NY: Cornell University Press, 2010.

–. "The Victorian Post-human: Transmission, Information and the Séance." In *The Ashgate Research Companion to Nineteenth-Century Spiritualism and the Occult*, ed. Tatiana Kontou and Sarah Willburn, 79–95. Surrey, UK: Ashgate, 2012.

Ghosh, Durba. "Decoding the Nameless: Gender, Subjectivity and Historical Methodologies in Reading the Archive of Colonial India." In *A New Imperial History*, ed. Kathleen Wilson, 297–316. Cambridge, UK: Cambridge University Press, 2004.

Gilman, Sander. *Difference and Pathology: Stereotypes of Sexuality, Race and Madness*. Ithaca, NY: Cornell University Press, 1985.

Goldsmith, Barbara. *Other Powers: The Age of Suffrage, Spiritualism and the Scandalous Victoria Woodhull*. New York: Alfred A. Knopf, 1998.

Gooding, David. "History in the Laboratory: Can We Tell What Really Went On?" In *The Development of the Laboratory: Essays on the Place of Experiment in Industrial Civilization*, ed. Frank A.J.L. James, 63–82. Houndmills, UK: Macmillan, 1989. http://dx.doi.org/10.1007/978-1-349-10606-6_5.

Gordon, Linda. *Heroes of Their Own Lives: The Politics and History of Family Violence, Boston, 1880–1960*. Urbana: University of Illinois Press, 2002.

Gorman, Daniel. *The Emergence of International Society in the 1920s*. Cambridge, UK: Cambridge University Press, 2012. http://dx.doi.org/10.1017/CBO9781139108584.

Green, H.A.V. "Foreword." In T.G. Hamilton, *Intention and Survival: Psychical Research Studies and the Bearing of Intentional Actions by Trance Personalities on the Problem of Human Survival*, ed. J.D. Hamilton, ix–xiv. Toronto: Macmillan, 1942.

–. "A Short Account of the Mary M. Photographs." *Light: A Journal of Psychical, Occult and Mystical Research* 49, 2521 (1929): 207–8.

Hacking, Ian. *Rewriting the Soul: Multiple Personality and the Sciences of Memory*. Princeton, NJ: Princeton University Press, 1995.

Haig-Brown, Celia. "Resistance and Renewal: First Nations Aboriginal Education in Canada." In *Race and Racialization: Essential Readings*, ed. Tania Das Gupta, Carl E. James, Roger C.A. Maaka, Grace-Edward Galabuzi, and Chris Andersen, 168–78. Toronto: Canadian Scholars' Press, 2007.

Hamilton, T.G. "The C.H. Spurgeon Case: Evidence Pointing to Survival and Continued Activity." *Light: A Journal of Psychical, Occult and Mystical Research* 53, 2752 (1933): 628–29.

–. *Intention and Survival: Psychical Research Studies and the Bearing of Intentional Actions by Trance Personalities on the Problem of Human Survival*. Ed. J.D. Hamilton. Toronto: Macmillan, 1942.

–. "Katie King Manifestations in the Mary M. Experiments." *Psychic Science: Quarterly Transactions of the British College* 11, 4 (1933): 254–78.

–. "Margery in Winnipeg: Three Séances of December 1926." In *Proceedings of the American Society for Psychical Research: The Margery Mediumship*, vol. 2, 556–67. New York: American Society for Psychical Research, 1933.

–. "The Mary M. Teleplasm of October 27th, 1929." *Psychic Science: Quarterly Transactions of the British College* 10, 4 (1932): 244–56.

–. "Some New Facts Regarding Teleplasms (an Address Delivered by Hamilton to Members and Friends of the British Medical Association on August 27th, 1930)." *Psychic Science: Quarterly Transactions of the British College* 9, 4 (1931): 262–70.

–. "A Study of the Winnipeg Group Mediumship in Its Relation to the Dawn Teleplasms." *Psychic Research* 29, 2 (1934): 117–30.

–. "Teleplasmic Phenomenon in Winnipeg." *Psychic Science: Quarterly Transactions of the British College* 8, 3 (1929): 179–208.

–. "Teleplasmic Phenomena in Winnipeg, Article 3: Mary M. Teleplasms, Including Another Miniature Face." *Psychic Science: Quarterly Transactions of the British College* 9, 2 (1930): 88–93.

Haraway, Donna. *Modest_Witness@Second_Millennium.FemaleMan©_Meets_OncoMouse™: Feminism and Technoscience*. New York: Routledge, 1997.

–. *Simians, Cyborgs and Women: The Reinvention of Nature*. New York: Routledge, 1991.

–. "Situated Knowledges: The Science Question in Feminism and the Privilege of Partial Perspective." In *The Gender of Science,* ed. Janet A. Kourany, 361–417. Upper Saddle River, NJ: Prentice Hall, 2002.

Harvey, John. *Photography and Spirit*. London: Reaktion Books, 2007.

Hazelgrove, Jenny. *Spiritualism and British Society between the Wars*. Manchester: Manchester University Press, 2000.

Helps, Lisa. "Body, Power, Desire: Mapping Canadian Body History." *Journal of Canadian Studies/Revue d'Etudes Canadiennes* 41, 1 (2007): 126–46.

Henderson, Linda Dalrymple. "Einstein and Twentieth-Century Art: A Romance of Many Dimensions." In *Einstein for the Twenty-first Century: His Legacy in Science, Art and Modern Culture,* ed. Peter L. Galison, Gerald Holten, and Silvan S. Schweber, 101–29. Princeton, NJ: Princeton University Press, 2008.

Heron, Craig. "The Workers Revolt, 1917-1925." In *Labouring Canada: Class, Gender, and Race in Canadian Working-Class History,* ed. Bryan D. Palmer and Joan Sangster, 138–56. Oxford: Oxford University Press, 2008.

Hersey, Merle W. *Seventy-Fifth Anniversary of the Lily Dale Assembly, 1879–1954, Lily Dale N.Y.: Condensed History*. Lily Dale, NY: Lily Dale Book Shop, 1954.

Hobbs, Margaret, and Joan Sangster. *The Woman Worker, 1926–29*. St. John's, NL: Canadian Committee on Labour History, 1999.

Howes, David. "Introduction." In *The Empire of the Senses: The Sensual Culture Reader,* ed. David Howes, 1–17. 2005. Reprint, Oxford: Berg, 2006.

–. "Introduction: The Revolving Sensorium." In *The Sixth Sense Reader,* ed. David Howes, 1–52. Oxford: Berg, 2009.

Huhndorf, Shari M. *Going Native: Indians in the American Cultural Imagination*. Ithaca, NY: Cornell University Press, 2001.

Hyslop, James H. "Signs of the Times." *Journal of the American Society for Psychical Research* 13, 2 (1919): 58–60.

–. "Spiritualism, Ignorance and Respectability." *Journal of the American Society for Psychical Research* 12, 12 (1918): 705–17.

Inden, Ronald. *Imagining India*. Oxford: Blackwell, 1992.

Iriye, Akira. *Global Community: The Role of International Organizations in the Making of the Contemporary World.* Berkeley: University of California Press, 2002.

Jacobus, Mary, Evelyn Fox Keller, and Sally Shuttleworth. "Introduction." In *Body Politics: Women and the Discourses of Science,* ed. Mary Jacobus, Evelyn Fox Keller, and Sally Shuttleworth, 1–10. New York: Routledge, 1990.

James, William. *The Will to Believe and Other Essays in Popular Philosophy*. New York: Green, 1896.

Jansen, Patricia. *Wild Things: Nature, Culture and Tourism in Ontario, 1790–1914*. Toronto: University of Toronto Press, 1995.

Joergenson, Joergen. *The Development of Logical Empiricism*. Vol. 2. Chicago: University of Chicago Press, 1951.

Jolly, Martyn. *Faces of the Living Dead: The Belief in Spirit Photography*. London: British Library, 2006.

Jones, Esyllt W. *Influenza 1918: Disease, Death and Struggle in Winnipeg*. Toronto: University of Toronto Press, 2007.

–. "Spectral Influenza: Winnipeg's Hamilton Family, Interwar Spiritualism and Pandemic Disease." In *Epidemic Encounters: Influenza, Society and Culture in Canada, 1918–20*, ed. Magda Fahrni and Esyllt Jones, 193–221. Vancouver: UBC Press, 2011.

Jung, Carl. *Modern Man in Search of a Soul*. Trans. W.S. Dell and Cary F. Baynes. New York: Harcourt, Brace and Company, 1933.

–, ed. *Studies in Word-Association: Experiments in the Diagnosis of Psychopathological Conditions Carried Out at the Psychiatric Clinic of the University of Zurich under the Direction of C.G. Jung, M.D., L.L.D.* Trans. M.D. Eder. New York: Moffat, Yard and Company, 1919.

Jütte, Robert. *A History of the Senses: From Antiquity to Cyberspace*. Cambridge, UK: Polity, 2005.

Kaempffert, Waldemar. "The Scientist." *Journal of the American Society for Psychical Research* 14, 10 (1920): 485–87.

Kaplan, Louis. *The Strange Case of William Mumler, Spirit Photographer*. Minneapolis: University of Minnesota Press, 2008.

Kasson, John F. *Houdini, Tarzan and the Perfect Man: The White Male Body and the Challenge of Modernity*. New York: Hill and Wang, 2001.

Kee, Kevin. *Revivalists: Marketing the Gospel in English Canada, 1884–1957*. Montreal and Kingston: McGill-Queen's University Press, 2006.

Kern, Stephen. *The Culture of Time and Space, 1880–1918*. Cambridge, MA: Harvard University Press, 1983.

King, Helen. "Once upon a Text: Hysteria from Hippocrates." In *Hysteria beyond Freud*, ed. Sander Gilman, Helen King, Roy Porter, and Elaine Showalter, 3–90. Berkeley: University of California Press, 1993.

King, Laura. *Family Men: Fatherhood and Masculinity in Britain, 1914–1960*. Oxford: Oxford University Press, 2015. http://dx.doi.org/10.1093/acprof:oso/9780199674909.001.0001.

King, Melanie. *The Dying Game: A Curious History of Death*. Oxford: Oneworld, 2008.

King, Richard. *Orientalism and Religion: Postcolonial Theory, India and 'The Mystic East.'* London: Routledge, 1999.

Klassen, Pamela. *Spirits of Protestantism: Medicine, Healing and Liberal Christianity*. Berkeley: University of California Press, 2011. http://dx.doi.org/10.1525/california/9780520244283.001.0001.

Kollar, Rene. *Searching for Raymond: Anglicanism, Spiritualism and Bereavement between the Two World Wars*. New York: Lexington Books, 2000.

Kontou, Tatiana, and Sarah Willburn, eds. *The Ashgate Research Companion to Nineteenth-Century Spiritualism and the Occult*. Surrey, UK: Ashgate, 2012.

Korsmeyer, Carolyn. *Making Sense of Taste: Food and Philosophy*. Ithaca, NY: Cornell University Press, 1999.

Kramer, Reinhold, and Tom Mitchell. *When the State Trembled: How A.J. Andrews and the Citizens' Committee Broke the Winnipeg General Strike*. Toronto: University of Toronto Press, 2010.

Kripal, Jeffrey J. *Authors of the Impossible: The Paranormal and the Sacred.* Chicago: University of Chicago Press, 2010. http://dx.doi.org/10.7208/chicago/9780226453897.001.0001.

Kucich, John J. *Ghostly Communion: Cross-Cultural Spiritualism in Nineteenth-Century American Literature.* Hanover, NH: Dartmouth College Press, 2004.

Lachapelle, Sofie. *Investigating the Supernatural: From Spiritism and Occultism to Psychical Research and Metaphysics in France, 1853–1931.* Baltimore, MD: John Hopkins University Press, 2011.

Lake, Marilyn, and Henry Reynolds. *Drawing the Global Colour Line: White Men's Countries and the International Challenge of Racial Equality.* Cambridge, UK: Cambridge University Press, 2008. http://dx.doi.org/10.1017/CBO9780511805363.

Lalvani, Suren. *Photography, Vision and the Production of Modern Bodies.* New York: State University of New York Press, 1996.

Laqua, Daniel. *Internationalism Reconfigured: Transnational Ideas and Movements between the World Wars.* New York: I.B. Tauris, 2011.

Latour, Bruno. *Pandora's Hope: Essays on the Reality of Science Studies.* Cambridge, MA: Harvard University Press, 1999.

–. *Reassembling the Social: An Introduction to Actor-Network Theory.* 2005. Reprint, Oxford: Oxford University Press, 2007.

–. *Science in Action: How to Follow Scientists and Engineers through Society.* Milton Keynes, UK: Open University Press, 1987.

Latour, Bruno, and Steven Woolgar. *Laboratory Life: The Construction of Scientific Facts.* London: Sage, 1979.

Lawless, Elaine J. *God's Peculiar People: Women's Voices and Folk Tradition in a Pentecostal Church.* Lexington: University Press of Kentucky, 1988.

LeBlanc, André. "The Origins of the Concept of Dissociation: Paul Janet, His Nephew Pierre, and the Problem of Post-hypnotic Suggestion." *History of Science* 39, 1 (2001): 57–66. http://dx.doi.org/10.1177/007327530103900103.

Lee, Erika. *At America's Gates: Chinese Immigration during the Exclusion Era, 1882–1943.* Chapel Hill: University of North Carolina Press, 2007.

Lee, Erika, and Judy Yung. *Angel Island: Immigrant Gateway to America.* Oxford: Oxford University Press, 2010.

Lefebvre, Henri. *The Production of Space.* Trans. Donald Nicholson-Smith. 1974. Reprint, Oxford: Blackwell, 2000.

Levine, Bruce. *Who Built America? Working People and the Nation's Economy, Politics, Culture and Society.* Vol. 2. Ed. American Social History Project Group. New York: Pantheon Books, 1992.

Lomas, David. *The Haunted Self: Surrealism, Psychoanalysis and Subjectivity.* New Haven, CT: Yale University Press, 2000.

Luckhurst, Roger. *The Invention of Telepathy, 1870–1901.* Oxford: Oxford University Press, 2002.

Lyons, Sherrie Lynn. *Species, Serpents, Spirits and Skulls: Science at the Margins in the Victorian Age.* Albany: State University of New York Press, 2009.

Mack, Phyllis. *Visionary Women: Ecstatic Prophecy in Seventeenth-Century England.* Berkeley: University of California Press, 1992.

Mahmood, Saba. "Agency, Performativity and the Feminist Subject." In *Pieties and Gender,* ed. Lene Sjorup and Hilda Romer Christensen, 11–45. Leiden: Brill, 2009. http://dx.doi.org/10.1163/ej.9789004178267.i-236.5.

Mar, Lisa Rose. *Brokering Belonging: Chinese in Canada's Exclusion Era, 1885–1995*. Toronto: University of Toronto Press, 2010. http://dx.doi.org/10.1093/acprof:oso/9780199733132.001.0001.

Marcus, Steven. *Freud and the Culture of Psychoanalysis*. New York: W.W. Norton and Company, 1987.

Marshall, David B. *Secularizing the Faith: Canadian Protestant Clergy and the Crisis of Belief, 1850–1940*. Toronto: University of Toronto Press, 1992.

McCann, Larry D., and Peter J. Smith. "Canada Becomes Urban: Cities and Urbanization in Historical Perspective." In *Canadian Cities in Transition*, ed. Trudi Bunting and Pierre Fillion, 69–99. Oxford: Oxford University Press, 1991.

McComas, Henry Clay. *Ghosts I Have Talked With*. Baltimore, MD: Williams and Wilkins Company, 1935.

McComas, Henry Clay, and Mary Winona McComas. *The McComas Saga: A Family History down to the Year 1950*. Baltimore, MD: McComas, 1950.

McDougall, William. "Further Observations on the 'Margery' Case." *Journal of the American Society for Psychical Research* 19, 6 (1925): 297–307.

McGarry, Molly. *Ghosts of Futures Past: Spiritualism and the Cultural Politics of Nineteenth-Century America*. Berkeley: University of California Press, 2008.

McLaren, Angus. *Reproduction by Design: Sex, Robots, Trees and Test-Tube Babies in Interwar Britain*. Chicago: University of Chicago Press, 2012. http://dx.doi.org/10.7208/chicago/9780226560717.001.0001.

–. *The Trials of Masculinity: Policing Sexual Boundaries, 1870–1930*. Chicago: University of Chicago Press, 1997. http://dx.doi.org/10.7208/chicago/9780226500690.001.0001.

McLuhan, Marshall. "Inside the Five Sense Sensorium." *Canadian Architect* 6 (1961): 49–54.

McMullin, Stan. *Anatomy of a Séance: A History of Spirit Communication in Central Canada*. Montreal and Kingston: McGill-Queen's University Press, 2004.

Meyer, Jessica. *Men of War: Masculinity and the First World War in Britain*. Hampshire, UK: Palgrave Macmillan, 2009. http://dx.doi.org/10.1007/978-0-230-30542-7.

Mitchinson, Wendy. *Giving Birth in Canada, 1900–1950*. Toronto: University of Toronto Press, 2002.

–. *The Nature of Their Bodies: Women and Their Doctors in Victorian Canada*. Toronto: University of Toronto Press, 1991.

Modern, John Lardas. *Secularism in Antebellum America*. Chicago: University of Chicago Press, 2011.

Monroe, John Warne. *Laboratories of Faith: Mesmerism, Spiritism, and Occultism in Modern France*. Ithaca, NY: Cornell University Press, 2008.

Moore, R. Laurence. *In Search of White Crows: Spiritualism, Parapsychology and American Culture*. New York: Oxford University Press, 1977.

Mosse, George L. *Fallen Soldiers: Reshaping the Memory of the World Wars*. New York: Oxford University Press, 1990.

Murphy, Gardner. "Dr. Prince and the Dora Case." In *Walter Franklin Prince: A Tribute to His Memory*, ed. Boston Society for Psychical Research, 66–81. Boston: Boston Society for Psychical Research, 1935.

Myers, Tamara. *Caught: Montreal's Modern Girls and the Law, 1869–1945*. Toronto: University of Toronto Press, 2006.

Nicholas, Jane. *The Modern Girl: Feminine Modernities, the Body and Commodities in the 1920s*. Toronto: University of Toronto Press, 2015.

Nicholson, Ian A.M. "Gordon Allport, Character and the 'Culture of Personality,' 1897–1937." In *Evolving Perspectives on the History of Psychology,* ed. Wade E. Pickren and Donald A. Dewsbury, 325–45. Washington, DC: American Psychological Association, 2002. http://dx.doi.org/10.1037/10421-016.

–. *Inventing Personality: Gordon Allport and the Science of Selfhood.* Washington, DC: American Psychological Association, 2003. http://dx.doi.org/10.1037/10514-000.

Nicolson, Malcolm. "The Introduction of Percussion and Stethoscopy to Early Nineteenth-Century Edinburgh." In *Medicine and the Five Senses,* ed. W.F. Bynum and Roy Porter, 134–53. Cambridge, UK: Cambridge University Press, 1993.

Nurse, Andrew. "Marius Barbeau and the Methodology of Salvage Ethnography in Canada, 1911-51." In *Historicizing Canadian Anthropology,* ed. Julia Harrison and Regina Darnell, 52–64. Vancouver: UBC Press, 2006.

O'Connell, Robert L. *Of Arms and Men: A History of War, Weapons and Aggression.* Oxford: Oxford University Press, 1989.

O'Hara Pincock, Jenny. *Trails of Truth.* Los Angeles, CA: Austin, 1930.

Opp, James. *The Lord for the Body: Religion, Medicine and Protestant Faith Healing in Canada, 1880–1930.* Montreal and Kingston: McGill-Queen's University Press, 2005.

–. "Picturing Communism: Yousuf Karsh, Canadair, and Cold War Advertising." In *The Cultural Work of Photography in Canada,* ed. Andrea Kunard and Carol Payne, 120–35. Montreal and Kingston: McGill-Queen's University Press, 2011.

Oppenheim, Janet. *The Other World: Spiritualism and Psychical Research in England, 1850–1914.* Cambridge, UK: Cambridge University Press, 1985.

Orsi, Robert. *Between Heaven and Earth: The Religious Worlds People Make and the Scholars Who Study Them.* Princeton, NJ: Princeton University Press, 2005.

Orvell, Miles. *The Real Thing: Imitation and Authenticity in American Culture, 1880–1940.* Chapel Hill: University of North Carolina Press, 1989.

Owen, Alex. *The Darkened Room: Women, Power and Spiritualism in Late Nineteenth Century England.* Chicago: University of Chicago Press, 1989.

–. *The Place of Enchantment: British Occultism and the Culture of the Modern.* Chicago: University of Chicago Press, 2004. http://dx.doi.org/10.7208/chicago/9780226642031.001.0001.

Palmer, Bryan D. *Working Class Experience: Rethinking the History of Canadian Labour, 1800–1991.* 2nd ed. Toronto: McClelland and Stewart, 1992.

Parr, Joy. *The Gender of Breadwinners: Women, Men and Change in Two Industrial Towns, 1880–1950.* Toronto: University of Toronto Press, 1990.

Patrias, Carmela, and Larry Savage. *Confrontation, Struggle and Transformation: Organized Labour in the St. Catharines Area.* Edmonton: Athabasca University Press, 2007.

Peterson, Christopher. *Kindred Specters: Death, Mourning and American Affinity.* Minneapolis: University of Minnesota Press, 2007.

Pickering, Andrew. *The Mangle of Practice: Time, Agency, and Science.* Chicago: University of Chicago Press, 1995. http://dx.doi.org/10.7208/chicago/9780226668253.001.0001.

Poovey, Mary. *A History of the Modern Fact: Problems of Knowledge in the Sciences of Wealth and Society.* Chicago: University of Chicago Press, 1998. http://dx.doi.org/10.7208/chicago/9780226675183.001.0001.

Prentice, Alison, Paula Bourne, Gail Cuthbert Brandt, Beth Light, Wendy Mitchinson, and Naomi Black. *Canadian Women: A History.* Toronto: Harcourt Brace Jovanovich, 1988.

Prince, Walter Franklin. "Some Highlights." *Boston Society for Psychic Research Bulletin: The "Walter"-"Kerwin" Thumb Prints,* 22 (April 1934): 74–85.

Randall, Ian M. *Evangelical Experiences: A Study in the Spirituality of English Evangelicism, 1918–1939*. Carlisle, UK: Paternoster, 1999.

Reiser, Stanley J. *Medicine and the Reign of Technology*. Cambridge, UK: Cambridge University Press, 1978.

–. "Technology and the Use of the Senses in Twentieth-Century Medicine." In *Medicine and the Five Senses,* ed. W.F. Bynum and Roy Porter, 262–73. Cambridge, UK: Cambridge University Press, 1993.

Rhine, J.B. *New Frontiers of the Mind: The Story of the Duke Experiments*. New York: Farrar and Rinehart, 1937.

Rhine, J.B., and Louisa E. Rhine. "One Evening's Observation on the Margery Mediumship." *Journal of Abnormal and Social Psychology* 21, 4 (1927): 401–21. http://dx.doi.org/10.1037/h0069024.

Rhine, Louisa E. *Mind over Matter*. New York: Macmillan, 1970.

Richards, Graham. "Britain on the Couch: The Popularization of Psychoanalysis in Britain, 1918-1940." *Science in Context* 13, 2 (2000): 183–230. http://dx.doi.org/10.1017/S0269889700003793.

Richardson, Mark A. "The Chinese Script and Cross-Correspondences." In *Proceedings of the American Society for Psychical Research: The Margery Mediumship,* vol. 2, 807–36. New York: American Society for Psychical Research, 1933.

Richardson, Sarah S. *Sex Itself: The Search for Male and Female in the Human Genome.* Chicago: University of Chicago Press, 2013.

Richet, Charles. *Thirty Years of Psychic Research: A Treatise on Metaphysics*. Trans. Stanley Debrath. Toronto: Macmillan, 1923.

Rieber, Robert W. *The Bifurcation of the Self: The History and Theory of Dissociation and Its Disorders*. New York: Springer, 2006. http://dx.doi.org/10.1007/0-387-27414-6.

Roberts, Mary Louise. *Civilization without Sexes: Reconstructing Gender in Postwar France, 1917–27*. Chicago: University of Chicago Press, 1994. http://dx.doi.org/10.7208/chicago/9780226721279.001.0001.

Robertson, Beth A. "Feminine Apparitions and Other Ghostly Teleplasm: Contesting and Constructing Womanliness in the Séances of Dr. T.G. Hamilton, 1918-1935." In *Spirit, Faith and Church: Women's Experiences in the English-Speaking World, 17th-21st Centuries,* ed. Laurence Lux-Sterritt and Claire Sorin, 143–67. Newcastle-upon-Tyne, UK: Cambridge Scholars, 2012.

–. "Radiant Healing: Gender, Belief, and Alternative Medicine in St. Catharines, Ontario, Canada 1927-1935." *Nova Religio: The Journal of Alternative and Emergent Religions* 18,1 (2014): 16–36.

–. "Spirits of Transnationalism: Gender, Race and Cross-Correspondence in Early Twentieth-Century North America and Europe." *Gender and History* 27, 1 (2015): 151–70. http://dx.doi.org/10.1111/1468-0424.12105.

Roediger, David R. *Towards the Abolition of Whiteness: Essays on Race, Politics and Working Class History*. New York: Verso, 1994.

–. *The Wages of Whiteness: Race and the Making of the American Working Class.* 1991. Reprint, London: Verso, 1996.

Rose, Nikolas. *Governing the Soul: The Shaping of the Private Self.* 2nd ed. London: Free Association Books, 1999.

–. *Inventing Our Selves: Psychology, Power and Personhood.* Cambridge: Cambridge University Press, 1998.

Said, Edward. *Orientalism*. New York: Vintage Books, 1979.

Sandell, Marie. "Regional versus International: Women's Activism and Organisational Spaces in the Inter-war Period." *International History Review* 33, 4 (2011): 607–25. http://dx.doi.org/10.1080/07075332.2011.620737.

Savage, Mike. *The Remaking of the British Working Class, 1840–1940*. London: Routledge, 1994.

Schiebinger, Londa L. *The Mind Has No Sex? Women in the Origins of Modern Science*. Cambridge, MA: Harvard University Press, 1989.

Schmidt, Leigh Eric. *Hearing Things: Religion, Illusion, and the American Enlightenment*. Cambridge, MA: Harvard University Press, 2000.

Sconce, Jeffrey. *Haunted Media: Electronic Presence from Telegraphy to Television*. Durham, NC: Duke University Press, 2000.

Scott, Clive. *The Spoken Image: Photography and Language*. London: Reaktion Books, 1999.

Sengoopta, Chandak. *The Most Secret Quintessence of Life: Sex, Glands and Hormones, 1850–1950*. Chicago: University of Chicago Press, 2006.

Shapin, Steven. "History of Science and Its Sociological Reconstructions." *History of Science* 20, 3 (1982): 157–211. http://dx.doi.org/10.1177/007327538202000301.

–. *A Social History of Truth: Civility and Science in Seventeenth-Century England*. Chicago: University of Chicago Press, 1994.

Shapin, Steven, and Christopher Lawrence. "Introduction: The Body of Knowledge." In *Science Incarnate: Historical Embodiments of Natural Knowledge*, ed. Steven Shapin and Christopher Lawrence, 1–20. Chicago: University of Chicago Press, 1998.

Shepherd, Ben. *A War of Nerves: Soldiers and Psychiatrists in the Twentieth Century*. Cambridge, MA: Harvard University Press, 2001.

Showalter, Elaine. *The Female Malady: Women, Madness and English Culture, 1830–1980*. New York: Pantheon Books, 1985.

–. "Hysteria, Feminism and Gender." In *Hysteria beyond Freud*, ed. Sander Gilman, Helen King, Roy Porter, and Elaine Showalter, 286–344. Berkeley: University of California Press, 1993.

Simmons, Christina. *Making Marriage Modern: Women's Sexuality from the Progressive Era to World War II*. Oxford: Oxford University Press, 2009. http://dx.doi.org/10.1093/acprof:oso/9780195064117.001.0001.

Sivulka, Juliann. *Soap, Sex and Cigarettes: A Cultural History of American Advertising*. 2nd ed. Boston: Wadsworth, 2012.

Skultans, Vielda. "Mediums, Controls and Eminent Men." In *Women's Religious Experience*, ed. Pat Holden, 15–26. London: Croom Helm, 1983.

Smajic, Srdjan. *Ghost-Seers, Detectives and Spiritualists: Theories of Vision in Victorian Literature and Science*. Cambridge, UK: Cambridge University Press, 2010. http://dx.doi.org/10.1017/CBO9780511712012.

Smith, Crosbie, and John Agar. "Introduction: Making Space for Science." In *Making Space for Science: Territorial Themes in the Shaping of Knowledge*, ed. Crosbie Smith and John Agar, 1–23. Houndmills, UK: Macmillan, 1998. http://dx.doi.org/10.1007/978-1-349-26324-0.

Smith, Mark M. *How Race Is Made: Slavery, Segregation and the Senses*. Chapel Hill: University of North Carolina Press, 2006.

–. *Sensing the Past: Seeing, Hearing, Smelling, Tasting and Touching in History*. Berkeley: University of California Press, 2007.

–. *Sensory History*. Oxford: Berg, 2007.

Spyer, Patricia. "Introduction." In *Border Fetishisms: Material Objects in Unstable Places,* ed. Patricia Spyer, 1–12. New York: Routledge, 1998.

Srigley, Katrina. *Breadwinning Daughters: Young Working Women in a Depression-Era City, 1929–1939*. Toronto: University of Toronto Press, 2010.

Stoler, Ann Laura. *Along the Archival Grain: Epistemic Anxieties and Colonial Common Sense*. Princeton, NJ: Princeton University Press, 2009.

Strong-Boag, Veronica. *The New Day Recalled: Lives of Girls and Women in English Canada, 1919–1939*. Toronto: Copp Clark Pittman, 1988.

Sturken, Marita, and Lisa Cartwright. *Practices of Looking: An Introduction to Visual Culture*. 2001. Reprint, Oxford: Oxford University Press, 2003.

Sudre, René. "Margery and Spiritism." *Journal of the American Society for Psychical Research* 20, 9 (1926): 524–31.

Tagg, John. *The Burden of Representation: Essays on Photographies and Histories*. London: Macmillan Education, 1988. http://dx.doi.org/10.1007/978-1-349-19355-4.

–. *The Disciplinary Frame: Photographic Truths and the Capture of Meaning*. Minneapolis: University of Minnesota Press, 2009.

Taves, Ann. *Fits, Trances and Visions: Experiencing Religion and Explaining Experience from Wesley to James*. Princeton, NJ: Princeton University Press, 1999.

Taylor, Eugene. *William James on Consciousness beyond the Margin*. Princeton, NJ: Princeton University Press, 1996.

Thorogood, Bracket K. "The 'Walter' Hands: A Study of Their Dermatoglyphics." In *Proceedings of the American Society for Psychical Research,* vol. 22, 79–88. New York: American Society for Psychical Research, 1933.

Thurs, David Patrick. *Science Talk: Changing Notions of Science in American Popular Culture*. New Brunswick, NJ: Rutgers University Press, 2007.

Thurschwell, Pamela. *Literature, Technology and Magical Thinking, 1880–1920*. Cambridge, UK: Cambridge University Press, 2001.

Tietze, Thomas R. *Margery*. New York: Harper and Row, 1973.

Tillyard, R.J. "R.J. Tillyard's Notes of His Séance with Margery." In *The Margery Mediumship: Record of Supernormal Production of Thumbprints,* ed. R.J. Tillyard, 6–12. New York: American Society for Psychical Research, 1931.

–. "Some Recent Personal Experiences with Margery." *Journal of the American Society for Psychical Research* 20, 12 (1926): 705–17.

Trachtenberg, Alan. *Shades of Hiawatha: Staging Indians, Making Americans*. New York: Hill and Wang, 2004.

Traverse, Lucy. "L'Âme Hu(main)e: Digital Effluvia, Vital Energies, and the Onanistic Occult." *Nineteenth-Century Contexts* 36, 5 (2014): 535–50. http://dx.doi.org/10.1080/08905495.2014.975395.

Tresch, John. *The Romantic Machine: Utopian Science and Technology after Napoleon*. Chicago: University of Chicago Press, 2012. http://dx.doi.org/10.7208/chicago/9780226812229.001.0001.

Trimm, Ryan S. "The Times of Whiteness; Or, Race between the Postmodern and the Postcolonial." In *Postcolonial Whiteness: A Critical Reader on Race and Empire,* ed. Alfred J. Lopez, 231–53. New York: State University of New York Press, 2005.

Tromp, Marlene. *Altered States: Sex, Nation, Drugs, and Self-Transformation in Victorian Spiritualism*. Albany: State University of New York Press, 2006.

Tucker, Jennifer. *Nature Exposed: Photography as Eye Witness in Victorian Science*. Baltimore, MD: John Hopkins University Press, 2005.

Urban, Hugh B. "'India's Darkest Heart': Kali in the Colonial Imagination." In *Encountering Kali: In the Margins, at the Center, in the West,* ed. Rachel Fell McDermott and Jeffrey J. Kripal, 169–89. Berkeley: University of California Press, 2003.

Vipond, Mary. *Listening In: The First Decade of Canadian Broadcasting, 1922–1932*. Montreal and Kingston: McGill-Queen's University Press, 1992.

Walker Bynum, Caroline. *Fragmentation and Redemption: Essays on Gender and the Human Body in Medieval Religion*. New York: Zone Books, 1991.

Walton, Daniel Day. "The McComas-Wood-Dunlap Commission and Its Report." In *Proceedings of the American Society for Psychical Research: The Margery Mediumship,* vol. 2. New York: American Society for Psychical Research, 1933.

–. "Notable Sittings in 1926: Houdini's Last Contact with the Case." In *Proceedings of the American Society for Psychical Research: The Margery Mediumship,* vol. 2, 514–39. New York: The American Society for Psychical Research, 1933.

Ward, Peter. *A History of Domestic Space: Privacy and the Canadian Home*. Vancouver: UBC Press, 1999.

Ward, Steven C. *Modernizing the Mind: Psychological Knowledge and the Remaking of Society*. Westport, CT: Praeger, 2002.

Warner, Marina. *Phantasmagoria: Spirit Visions, Metaphors, and Media into the Twenty-First Century*. Oxford: Oxford University Press, 2006.

Watts, Ruth. *Women and Science: A Social and Cultural History*. London: Routledge, 2007.

Weinbaum, Alys Eve, Lynn M. Thomas, Priti Ramamurthy, Uta G. Poiger, Madeleine Yue Dong, and Tani E. Barlow. "The Modern Girl as Heuristic Device: Collaboration, Connective Comparison, and Multidirectional Citation." In *The Modern Girl around the World: Consumption, Modernity, and Globalization,* ed. Alys Eve Weinbaum, Lynn M. Thomas, Priti Ramamurthy, Uta G. Poiger, Madeleine Yue Dong, and Tani E. Barlow, 1–24. Durham, NC: Duke University Press, 2008. http://dx.doi.org/10.1215/9780822389194.

Westfall, William. *Two Worlds: The Protestant Culture of Nineteenth-Century Ontario*. Montreal and Kingston: McGill-Queen's University Press, 1989.

Winter, Alison. "A Calculus of Suffering: Ada Lovelace and the Bodily Constraints on Women's Knowledge in Early Victorian England." In *Science Incarnate: Historical Embodiments of Natural Knowledge,* ed. Steven Shapin and Christopher Lawrence, 202–39. Chicago: University of Chicago Press, 1998.

–. *Mesmerized: Powers of Mind in Victorian Britain*. Chicago: University of Chicago Press, 1998.

Wood, W.R. "Science and Survival: A Canadian Investigator's Examination of the Evidence." *Light: A Journal of Psychical, Occult and Mystical Research* 56, 2921 (1936): 833–34.

Worcester, Elwood. "Dr. Prince as a Psychical Researcher and as a Psychiatrist." In *Walter Franklin Prince: A Tribute to His Memory,* ed. Boston Society for Psychical Research, 9–16. Boston: Boston Society for Psychical Research, 1935.

Yoshihara, Mari. *Embracing the East: White Women and American Orientalism*. Oxford: Oxford University Press, 2003.

Zaidi, Waqar H. "'Aviation Will Either Destroy or Save Our Civilization': Proposals for the International Control of Aviation, 1920–1945." *Journal of Contemporary History* 46, 1 (2011): 150–78. http://dx.doi.org/10.1177/0022009410375257.

Photo Credits

1 Hamilton Family Fonds, PC12 (A.79-41), box 9, folder 4, "Telekinesis," University of Manitoba Archives and Special Collections, Winnipeg.
2 Hamilton Family Fonds, PC12 (A.79-41), box 9, folder 1, "Telekinesis," University of Manitoba Archives and Special Collections, Winnipeg.
3 T.G. Hamilton, "Levitations by Psychic Force (Telekinesis)," Hamilton Family Fonds, PC12 (A.79-41), box 8, folder 1, University of Manitoba Archives and Special Collections, Winnipeg.
4 Hamilton Family Fonds, PC12 (A.79-41), box 8, folder 5, University of Manitoba Archives and Special Collections, Winnipeg.
5 Hamilton Family Fonds, MSS14 (A.79-41), box 9, folder 13, University of Manitoba Archives and Special Collections, Winnipeg.
6 Hamilton Family Fonds, MSS14 (A.79-41), box 9, folder 12, University of Manitoba Archives and Special Collections, Winnipeg.
7 Hamilton Family Fonds, MSS14 (A.79-41), box 9, folder 13, University of Manitoba Archives and Special Collections, Winnipeg.
8 Hamilton Family Fonds, PC12 (A.79-41), box 8, folder 1, University of Manitoba Archives and Special Collections, Winnipeg.
9 Maines Pincock Family Fonds, GA64, series 8, file 182, University of Waterloo Library, Waterloo.
10 Hamilton Family Fonds, PC12 (A.79-41), box 9, folder 28e, University of Manitoba Archives and Special Collections, Winnipeg.

11 Hamilton Family Fonds, PC12 (A.79-41), box 10, folder 52c, University of Manitoba Archives and Special Collections, Winnipeg.

12 Society for Psychical Research Fonds, MS39/14, Cambridge University Library, London, England.

13 Hamilton Family Fonds, PC12 (A.79-41), box 9, folder 20d, University of Manitoba Archives and Special Collections, Winnipeg.

14 Society for Psychical Research Fonds, MS39/14, Cambridge University Library, London, England.

Index